RESTAURANT SERVICE *basics*

SECOND EDITION

Sondra J. Dahmer

Kurt W. Kahl

JOHN WILEY & SONS, INC.

This book is printed on acid-free paper. ⊗

Published by John Wiley & Sons, Inc., Hoboken, New Jersey.

Published simultaneously in Canada.

For general information on our other products and services, or technical support, please contact our Customer Care Department within the United States at 800-762-2974, outside the United States at 317-572-3993 or fax 317-572-4002.

Wiley also publishes its books in a variety of electronic formats. Some content that appears in print may not be available in electronic books.

For more information about Wiley products, visit our Web site at http://www.wiley.com.

Library of Congress Cataloging-in-Publication Data:

Dahmer, Sondra J.
Restaurant service basics / Sondra J. Dahmer, Kurt W. Kahl. – 2nd ed.
p. cm.
Includes index.
ISBN 978-0-470-10785-0 (pbk. : alk. paper)
1. Table service. 2. Waiters. 3. Waitresses. I. Kahl, Kurt W. II. Title.
TX925.D33 2009
642′.6–dc22 2008002721

Printed in the United States of America

10 9 8 7 6 5 4 3 2 1

Contents

Chapter 2: Types of Establishments, Types of Service, and Table Settings 17

Chapter 3: Before the Guests Arrive 35

Chapter 4: Initiating the Service 61

Chapter 5: Serving the Meal 79

Chapter 7: Handling Service Using Technology 115

Chapter 8: Wine and Bar Service 137

Preface

Restaurant Service Basics, 2nd edition, is a practical guide for those who want to learn the core skills of professional table service in restaurants. Actual and prospective servers, as well as managers, supervisors, and teachers who train servers, will find this an invaluable resource for classroom use, restaurant training, or self-training.

This book discusses the server's job, types of establishments, and different types of service, including French, Russian, English, American, banquet, family-style, buffets, and more. Current issues such as embracing diversity, preventing harassment, and maintaining a drug-free workplace are also discussed. The text walks the reader through the dining experience from taking reservations, preparing the dining room, and greeting and serving guests to presentation of the check, and instructs the server on how to troubleshoot potential problems that may occur along the way. Safety, sanitation, and medical emergencies are addressed. Current information on ever-changing restaurant technology has a chapter of its own. The final chapter covers alcoholic beverage service, with all of its ramifications to the restaurant business.

Restaurant Service Basics, 2nd edition, will enable readers to develop the consummate service skills required to handle all phases of providing quality service, increasing their tips, and capturing repeat business for the restaurant. This is a great training tool for new servers and a reference tool for veteran servers. Servers can learn the techniques of serving that will perfect their job performance and guarantee success.

Note to Teachers and Trainers

This text is intended for use in a classroom or for training servers employed by a restaurant. Assigning chapters, key terms, review questions, or relevant projects from the text will help trainees gain a firm grasp of the fundamentals of providing competent service to guests to ensure their guests have a pleasant dining experience. You can also help your trainees use the case at the end of each chapter to build competencies through group discussion. Additional resources at the end of the book provide definitions of key terms, pronunciations of menu terms, and information for further reading. The key terms, review questions, projects, cases, and exam can all be used to evaluate the trainee's knowledge of serving procedures.

An ***Instructor's Manual*** (ISBN 978-0-470-25736-4) can be obtained by contacting your Wiley sales representative. If you don't know who your representative is, please visit **www.wiley.com/college** and click on "Who's My Rep?" An electronic version of the ***Instructor's Manual*** is available to qualified instructors on the companion Web site, at **www.wiley.com/college/dahmer**.

Note to the Server or Trainee

You can use this manual as a text or self-training manual to help you learn how to serve competently or increase your serving knowledge and skills. After reading the chapters, you should define the terms and answer the review questions to gauge your understanding of the key concepts discussed and complete as many of the projects and cases as possible to review proper serving procedures. Use Resource B to look up menu terms.

The opportunity for employment as a server in a restaurant is favorable and can be a lifetime occupation. The challenge is great, but with hard work and a determination to succeed, you will be rewarded not only financially, but also with a sense of accomplishment.

Web site links, chapter objectives, and definitions of key terms are available at the student companion Web site, at **www.wiley.com/dahmer**.

New to This Edition

This edition provides updated material about the server's job and the trends and practices that have emerged since the first edition, such as:

- An updated description of a server's job
- A comparison between service and hospitality
- A detailed discussion about types of establishments
- New and updated information on the role of technology in taking reservations
- Updated information about food allergies and food trends
- A discussion about coffee and tea service

- A presentation of the most up-to-date food safety and sanitation guidelines based on the most recent guidelines from the Centers for Disease Control and Prevention (CDC) and the Food and Drug Administration (FDA) Model Food Code
- A detailed discussion of emergency procedures for various situations, including severe weather
- The most current first-aid procedure for conscious choking emergencies, provided by the American Red Cross
- An expanded and updated chapter on technology in restaurants, including handheld order terminals, reservation management and seating software, and guest paging devices
- A discussion about the responsibilities associated with serving alcohol in restaurants
- A new ***Instructor's Manual*** (ISBN 978-0-470-25736-4), which includes lecture notes, final exam and answer key, chapter quizzes, and additional activities.

Specific changes to this edition include the following:

Learning Objectives have been included to provide the reader with a road map of the key points covered in each chapter. They help highlight what the reader should be particularly focused on throughout each chapter.

Bolded Key Terms are included within each chapter and defined as they appear. They are then listed at the end of each chapter in the order in which they appear. This list indicates the terms the reader should understand from reading each chapter.

The **Review** at the end of each chapter can be used to reexamine the chapter content. The questions cover the salient points in the chapter material. Review questions can drive discussion in class or be assigned for homework.

The **Projects** at the end of each chapter are one of the greatest strengths of the text because they are applied in nature. The projects have been updated to address current issues servers encounter in today's foodservice industry.

The **Mini-Cases** included at the end of the ***first edition*** are now included at the end of each chapter. These scenarios have been updated to reflect current issues encountered at a foodservice establishment, such as handling harassment, addressing a problem at the buffet, handling a reservation error, reacting to a guest complaint, and addressing an emergency situation. These mini-cases are intended to help students apply what they have learned in a particular chapter to situations they might encounter as a server in a restaurant operation. Each case includes a set of questions that instructors can either assign as homework or use to engage students in the classroom and drive discussion.

The **photos and drawings** have been updated to more effectively illustrate key concepts, reflect the current restaurant environment, and add to the visual appeal of this edition.

Additional materials include a new Resource A: Definition of Key Terms from the Text; Resource B: Definition of Menu and Service Terms; and Resource C: Recommended Resources for Further Information.

Acknowledgments

Our goal with this revision of ***Restaurant Service Basics*** is to provide our readers with an accessible, reader-friendly guidebook on the fundamental skills required to provide guests with the best service possible. We could not have completed this edition of ***Restaurant Service Basics*** without the help, directly and indirectly, of many people. We are grateful to Cindy Rhoads, our developmental editor, for her excellent advice and encouragement to bring our manuscript through to completion.

We would also like to acknowledge the help of our reviewers, who carefully reviewed ***Restaurant Service Basics*** and suggested updates for our second edition to ensure it provides the most up-to-date and accurate information about the server's valuable role in the overall restaurant operation:

Natasa Christodonlidou, University of Nevada, Las Vegas

Lisa Clark, Cypress College, CA

James Feerasta, University of Akron, OH

Henry L. Jancose, University of Houston, Conrad N. Hilton College of Hotel and Restaurant Management, TX

Cindy A. Komarinski, Westmoreland County Community College, PA

Madoka Watabe-Belzel, California Polytechnic State University

Ronald Wolf, Florida Community College at Jacksonville

Among other people who helped us, we want to thank the staff of many fine eateries who took the time to share experiences, answer questions, and pose for pictures. And thank you also to our families for their advice and support while we worked on revising this book.

Chapter One

The Server

Objectives

After reading this chapter, you will be able to:

- ✔ Define the job of a server.
- ✔ Distinguish between service and hospitality.
- ✔ List the advantages of the job as server.
- ✔ Understand the job qualifications.
- ✔ Realize the importance of good personal appearance.
- ✔ Understand how a server fits into the restaurant organization.
- ✔ Understand the importance of getting along with coworkers.
- ✔ Value diversity in guests and coworkers.
- ✔ Handle harassment on the job.
- ✔ Keep violence and drugs out of the workplace.
- ✔ Understand the importance of safety and sanitation in a restaurant.

Servers, also referred to as **waitstaff**, are restaurant employees who create a dining experience for guests in a restaurant by making them feel welcome and comfortable, taking their orders, serving the meal, clearing the area, and setting the table for the next party of guests. Servers also maintain the service areas of the dining room and the kitchen so that everything is ready for smooth, efficient service (see Job Description for a Server, Figure 1-1, and www.nraef.org for descriptions of related jobs). This chapter discusses service and hospitality, the advantages of a server's job, the qualifications necessary to become a server, the importance of the server's personal appearance on the job, the server's position in the organizational structure of the restaurant, and the server's role within a team of fellow employees and supervisors.

Although the server's tasks may seem clear, many aspects of the job involve issues of great concern to the guest and the establishment, as well as the server. In this chapter, we will discuss issues that include recognizing the diversity of coworkers and guests, preventing harassment on the job, deterring violence and drugs in the workplace, and practicing restaurant safety and sanitation. Some of these issues are discussed in greater detail later in this book.

Service and Hospitality

While a close relationship exists between service and hospitality, there is a distinct difference between the two. **Service** is the act of filling the needs, wants, and desires of the guests. Service is what servers provide to meet the expectations of the guests when they come to dine. Guests expect a clean table, clean dishes and utensils, safe food, hot foods served hot, and cold foods served cold.

Hospitality goes beyond the service guests expect of servers. **Hospitality** means creating a pleasant dining experience for your guests with small gestures like giving a friendly greeting, smiling—even when very tired, remembering names, hanging up coats, pulling out chairs, remembering a returning guest's favorite drink, knowing exactly what is ordered, and anticipating what the guest needs next. It is paying close attention to detail. It is acknowledging guests promptly, making friendly small talk, and saying a pleasant goodbye when the guests leave the restaurant. It is reading the guests and always making them feel comfortable, welcome, and important. Hospitality is a key element to bringing guests back and to increasing the amount of your tip.

Advantages of Being a Server

Despite tired feet, unruly guests, and job demands, there are many advantages to holding a job as a server. These include:

- Monetary benefits
- Flexible hours

JOB DESCRIPTION FOR SERVER

PREPARE RESTAURANT FOR ... by
Cleaning tables in an assigned area of the restaurant
Arranging the china, silverware, napkins, glassware, and centerpieces
Setting up and maintaining serving area

MAINTAIN FOOD AND RESTAURANT SAFETY AND SANITATION by
Adhering to ... of food and utensils according to Hazard Analysis and Critical Control Point (HACCP) guidelines
Attending to accident prevention
Being prepared for unforeseen accidents and emergencies

... DINING EXPERIENCE FOR GUESTS by
Making guests feel welcome, comfortable, and important
... of hospitality
Working as a team member with coworkers for smooth restaurant operation

HELP GUESTS MAKE FOOD AND BEVERAGE SELECTIONS by
Presenting the menus
... food and beverages
Offering cocktails and suggesting wine
...

SUBMIT ... KITCHEN AND BAR by
Taking the ...
... requests
... kitchen and bar
Timing the meal

SERVE THE MEAL by
Staging meal according to ... service
Serving beverages, cocktails, or beer
Opening and serving wines
Picking up, garnishing, and serving ...
Providing ...

MONITOR THE MEAL IN PROGRESS by
Inquiring of needs
Replenishing items
Refilling ... ordering more ...
Removing completed courses
Adhering to alcohol control policies

CONCLUDE THE DINING EXPERIENCE by
Totaling charges and presenting the guest check
Accepting and settling payment
Taking leave of guests and inviting them back again

FIGURE 1-1 Sample Job Description for a Server: Management looks for employees who have the qualifications to become good servers and perform the tasks required of them.

- Contact with people
- Minimal investment in wardrobe
- Pleasant surroundings
- Job satisfaction

One of the advantages of being a server is that you may be compensated well for providing good services to the guests you serve in a restaurant. In elegant restaurants and restaurants with quick turnover, a server who provides good service can make more money in tips and wages than a cook, administrative assistant, police officer, flight attendant, or schoolteacher. You may also be eligible for free or reduced-priced meals from the restaurant.

Restaurant establishments operate for several hours extending over the periods for breakfast, lunch, dinner, and into the evening. You might consider working weekends and evenings a disadvantage, but the flexible hours allow many people to work around another job or family responsibilities.

As a server, another advantage is that you can meet many interesting people as you serve your guests. You have the chance to interact with people of all ages and from all walks of life. Occasionally, you may even serve a celebrity.

Unlike a job in an office, which requires a large investment in business clothing, as a server, you are only required to buy a few uniforms and comfortable shoes.

Some people also find working as a server in pleasant surroundings advantageous. Many restaurants have a very elaborate decor and atmosphere. You may also have an opportunity to sample a variety of food.

Finally, you can gain personal satisfaction from doing a job exceptionally well and making people happy. Compliments and tips from the guests and words of praise from your manager are your rewards for providing good service. Take pride in your job. Waiting tables can be a rewarding part-time position or a full-time career that provides you with a very good living. The skills and knowledge you acquire will translate to all aspects of your life.

Qualifications for the Job

Management looks for employees who are:

- Reliable
- Cooperative
- Personable
- Healthy
- Clean and neat

- Knowledgeable
- Persuasive
- Attentive
- Diplomatic
- Good managers

People who have these qualifications make good servers. To qualify for a waitstaff position, you must be:

1. *Reliable.* Management must feel confident that you are capable of fulfilling your job responsibilities. You must report to work on time, serve the guests properly, and complete all tasks assigned to you.
2. *Cooperative.* You must have a good attitude. You must be willing to work hard with your coworkers to complete the work assigned. If you are a good team player, and strive to make your team successful, you will be successful at your job. Being cooperative also means adapting to the policies of management. You should work hard and learn and stay focused on your job even under pressure and time restraints.
3. *Personable.* A server is chosen for his or her pleasant personality, comfort around people, ability to make small talk, and social skills. Take the initiative to be friendly, patient, and courteous to guests, coworkers, and management, and never be rude. A server should have a good sense of humor. However, never be familiar with guests or use terms such as "folks," "honey," or "you guys" when addressing guests.
4. *Healthy.* Because a serving job requires the server to be in close contact with guests, coworkers, and food, you must maintain good health. Staying healthy helps you to avoid spreading disease and to maintain a good appearance. A healthy server looks good, performs well on the job, and is able to lift and carry heavy trays.
5. *Clean and neat.* A server may be one of the only restaurant employees that the guest sees and must present a tidy appearance that reflects the image of a clean and neat restaurant. Particular attention should be given to hair, nails, uniform, and shoes. A server should be clean and neat in both appearance and with respect to handling food and serviceware in the restaurant.
6. *Knowledgeable.* A good server must know the appropriate methods for serving tables and how to apply them in a seamless and efficient manner. You must have an extensive knowledge of the menu so that you can answer guests' questions intelligently, suggest foods, and help increase sales. To complete the guest's order and total the guest check, you must know basic math and have simple

computer skills. Make note of local history, events, and cuisine trends so you can make conversation with guests.

7. *Persuasive*. You must have the ability to sell yourself, the restaurant, and menu items as you serve your guests. To do this, you must be able to communicate well. Your persuasive talent ensures that the guest's wants and needs are met, and **suggestive selling**—that is, suggesting additional food items—adds to the pleasure of the meal, as well as increases the size of the check, the restaurant profits, and your tip.
8. *Attentive*. Guests may need their server at any time during the meal, so a good server always concentrates on the job and never leaves a station unattended for long. You must be aware of the progress of the meal at each table and anticipate needs as they arise. When guests toy with an empty glass, gaze into space, or peek at their watches, they are sending signals that they need something. If a guest tastes the food, puts down his or her fork, and pushes the plate away, something is wrong and servers need to remedy the situation. Refill empty water glasses and coffee cups, but do not hover to an excessive degree. Take pride in the appearance of the dining room by keeping it orderly as you work.
9. *Diplomatic*. Servers should handle complaints in a diplomatic way to assure the guest is satisfied and will come back again. A good server must be thick-skinned and not take criticism personally.
10. *A good manager*. You must know how to manage your time well. Prepare your area in advance, do the most important tasks first, multitask so you address your guests' every need, and do so in a quiet, controlled manner that is not interruptive to the dining room.

Personal Appearance on the Job

Your appearance as a server on the job gives guests their first impression of you and, consequently, a lasting impression of the restaurant. Because you are one of the few members of the restaurant staff a guest sees, a guest may judge the restaurant largely on your appearance and service.

THE UNIFORM

A **uniform** is a garment that identifies the occupation of the wearer. Nurses, police officers, pilots, and members of the armed service, as well as servers, wear uniforms. The appearance of your uniform leaves an impression on your guests (Figure 1-2). A clean and neat uniform reflects an image of a sanitary restaurant. If your uniform is soiled or wrinkled, you will not impress the guests favorably, and the guests,

FIGURE 1-2 Server in Uniform: You are the main restaurant employee the guest sees so be sure your uniform is clean and neat. Your appearance can help create a good impression for the entire operation.
Courtesy of PhotoDisc/Getty Images

deciding that your uniform reflects the standards of the whole operation, may never return.

A server should wear a clean uniform each workday and keep an extra one at work in case of emergency. Skirts, pants, jackets, and ties should be neat, clean, and pressed. Most uniforms today are made of synthetic fiber blends that are easy to maintain. If you do spill food on a uniform, remove the stains as soon as possible and launder the garment according to the manufacturer's directions. Uniforms in disrepair are as unacceptable as soiled uniforms. Repair torn hems and seams, and replace buttons before you wear the uniform again.

A uniform should fit well; if it is too small, it restricts movement. Pants should fit smoothly and not be too tight. If a female server wears a skirt as part of a uniform, it should have a fashionable hem length, but not be so short that she feels conspicuous when reaching or bending.

Shoes are part of the uniform and should receive daily attention. Buy sturdy shoes with closed toes, low heels, and arch supports. Shoes should have rubber soles to minimize slips and falls. Replace shoes or have worn heels and soles repaired, and be sure your shoes are clean and polished for work. Have a second pair of shoes at work, especially if you are a full-time server, and change your shoes occasionally to prevent foot and back problems.

Female servers might consider wearing support hose for comfort and pantyhose for good appearance, because the job requires reaching and bending. Keep an extra pair in your locker or purse in case of a run.

Wedding and engagement rings and classic watches may be worn, but decorative jewelry, such as bracelets, dinner rings, and lapel pins are not appropriate as part of a uniform. Decorative jewelry does not look professional and is not sanitary when you are working with food.

HYGIENE AND GROOMING

Because you are working with the public, careful attention must be given to your personal hygiene and grooming. **Hygiene** means practices that promote personal cleanliness and good health, and **grooming** means the process of making your appearance neat and attractive. For proper appearance and to look well physically, you must have the proper amount of rest each night. Bathe daily, and apply an antiperspirant to prevent body odors. Brush your teeth, use a mouthwash, and see a dentist twice a year. Use breath mints or breath sprays at work. Never smoke or chew gum in front of guests.

Wear your hair in a simple, stylish manner pulled back from your face, and avoid extreme hairstyles. Be sure your hair is clean and combed. Use effective hair restraints, such as caps, ponytail bands, headbands, barrettes, and other accessories designed to be part of the uniform, to prevent the contamination of food or food contact surfaces.

Servers should be sure their hands and nails are clean, because they are on display and touching food and utensils. Scrub your nails, and trim them to a short, even length. Female servers may wear a conservative color or clear nail polish. Keep your hands away from your hair and face. Wash your hands thoroughly with soap after using the restroom, clearing soiled dishes, or handling money.

Male servers should be clean shaven. Female servers should use a minimum amount of makeup, such as a conservative application of eye makeup and lipstick. Perfumes and colognes do not enhance food aromas and should not be worn.

Check your total appearance in a mirror before you start work. Ask yourself, "If I owned a restaurant, would I want me as an employee?"

Your Role in the Restaurant Organization

The goals of a restaurant organization are to satisfy the guests and make a profit. To reach these goals, employees should work together in a united effort. Employees are expected to perform their jobs and to support coworkers in a team approach. This bonding of efforts will give the guests the best service.

Employees are given a job description and assigned activities so that no work is duplicated or omitted. A server's job is to assist management by giving good service, being efficient, avoiding waste, attending to safety and sanitation, and following the rules and regulations set by management. Remember, everyone will have job longevity if the restaurant employees work as a team to please the guests and guests return time after time.

The organization or arrangement of jobs in a restaurant diagrammed on paper is known as the **organization chart**. Knowing how you fit into the organization chart helps you function in your job. In the dining room organizational structure, **bussers** and servers report to their immediate supervisor, who may be a **head server**. The head server reports to the **host** or **maître d'hôtel,** who reports to the **dining room manager**. The dining room manager reports to management. In small restaurants, one person may assume several of these positions. For instance, your immediate supervisor may be a maître d'hôtel or host who also assumes the head server's job. In large restaurants, additional staff, such as an **expeditor,** may help the server bring food from the kitchen. Figure 1-3 shows a traditional organization chart in a restaurant.

To avoid causing problems within the organization, follow the organizational structure. When you have questions or problems concerning the rules, regulations, or policies, talk to your immediate supervisor. Your supervisor has more experience and is in a position to solve certain problems.

Teamwork with Coworkers and Supervisors

Teamwork means cooperating and working together with coworkers and the supervisor of the dining room to serve the public. A serving team is like a football team, working toward a common goal. Whether the goal is a touchdown or a satisfied guest, the principles are the same. Use the following guidelines to help foster teamwork:

- *Arrive to work with a positive attitude.* Leave your personal problems at home, and do not discuss them with coworkers or guests. Be cheerful and happy in your work; this attitude spreads to other workers and to the guests. Remember, guests come to the restaurant to relax and enjoy a special occasion in pleasant surroundings. If you have problems on the job, work them out or discuss them with your supervisor instead of with coworkers. Work can be pleasant or unpleasant, depending on your attitude toward it.

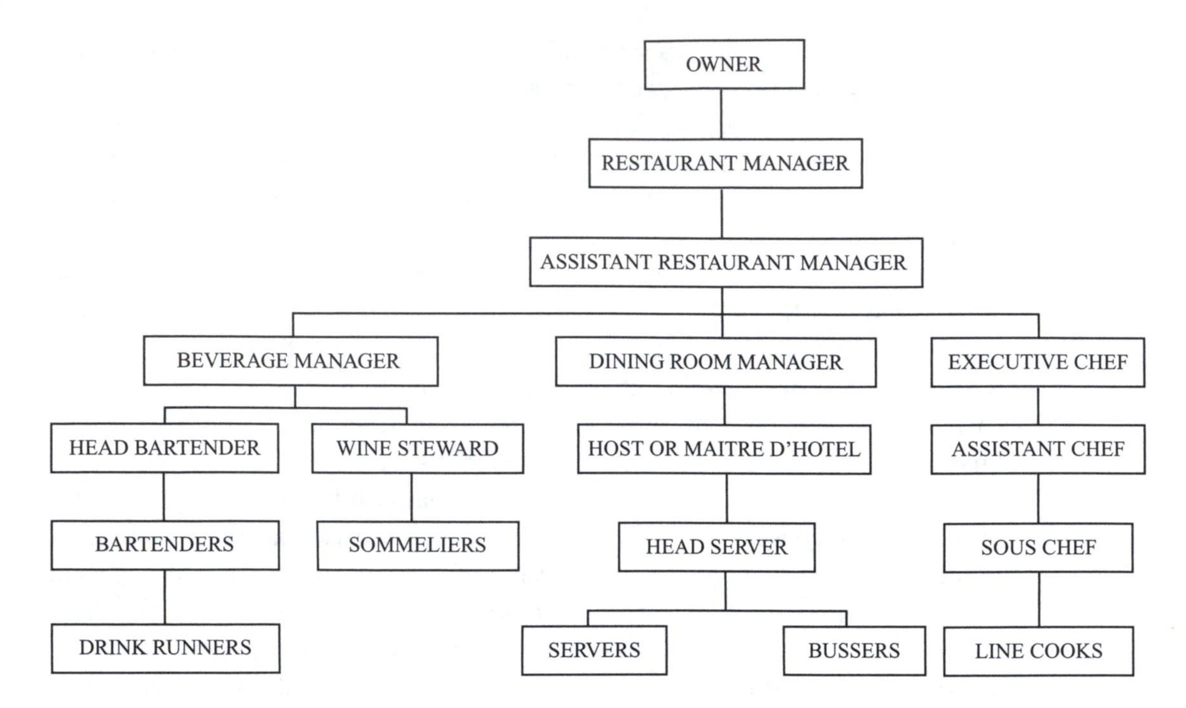

Restaurant Management: Ensures that the restaurant operates efficiently and profitably.

Dining Room Manager: Directs and coordinates foodservice in the dining room; hires, trains, and supervises employees; handles budgets, payroll, and purchasing.

Host or Maître d'Hôtel: Schedules shifts and assigns stations; holds daily meetings with staff; controls flow of seating; greets, seats, and provides menus to guests.

Head Server: Supervises and coordinates dining room employees for a section of the dining room; may greet and seat guests; may serve guests.

Busser: Assists server in serving water, bread and butter, and coffee refills; clears soiled dinnerware; resets table.

FIGURE 1-3 Dining Room Organization Chart Including a Brief Definition of Service Jobs: The relationship of the personnel in a traditional dining room is shown in this organization chart.

- *Avoid raising your voice to any coworker or the chef when problems arise.* Instead, try to handle the situation calmly. Some managers will train you in other jobs (chef, busser, and host) or allow you to experience other positions to build empathy with coworkers. Remember, you are a professional and a representative of the restaurant, and you should handle problems with coworkers in a professional manner.
- *Give coworkers assistance when they need help and you are not busy.* For example, help them carry trays of food when they are serving an especially large party. If a guest in a coworker's station asks you for service, either cheerfully render

the service or inform the guest's server. As a member of the team, the coworker should return the favor when you are busy.

- *Refrain from chatting or gossiping with coworkers in the dining room.* Your responsibility lies with your guests when you are on duty. If you have spare time, use it productively to check your station, clean and fill condiment containers, fold napkins, and replenish the sidestand.
- *If you are ill, notify your supervisor as soon as possible.* Absenteeism without proper notification may mean that a coworker must assume double duty if a replacement is unobtainable.

If even one member of your restaurant's team falls short of his or her duties, it creates a ripple effect and puts pressure on everyone. The outcome affects the entire operation.

Issues Regarding Restaurant Employment

Current issues in restaurant employment include diversity, harassment, violence and drugs, safety, and sanitation. Some of these issues are covered more thoroughly in other chapters, but they deserve mention here.

RESPECTING THE DIVERSITY OF ALL PEOPLE

Diversity means difference or unlikeness and refers to the fact that each person is unique with regard to race, color, creed, ethnicity, religion, national origin, gender, sexual orientation, disability, age, marital status, socioeconomic status, veteran status, belief, or ideology, to name just a few dimensions. Diversity is reflected in the way each person walks, talks, thinks, and behaves. The concept of diversity encompasses accepting, respecting, and embracing the rich dimensions of diversity contained within each individual. Your challenge as a server is to recognize, appreciate, value, and respect the differences you encounter in each person, whether you are interacting with coworkers or guests.

The law states that every person—regardless of such things as their race, religion, or nationality—deserves to work in a safe and pleasant environment. If you have intolerance for coworkers' differences, you will undermine the team effort necessary to serve the public well. Likewise, all guests should receive equal treatment regardless of their diverse backgrounds and varying cultural behaviors. As a server, it is imperative that you respect these differences. Your conduct as a server must be tolerant and respectful.

PREVENTING HARASSMENT ON THE JOB

Harassment is to disturb, worry, unnerve, or torment by continuous small attacks. All employees have the legal right to a work environment free of verbal or physical harassment. One form of harassment, sexual harassment, is defined as unwelcome sexual advances, requests for sexual favors, and other verbal or physical conduct of a sexual nature. Sexual harassment violates employee rights, creates stress, and reduces productivity. If you are sexually harassed on the job, follow these guidelines:

1. Explain to the person doing the harassing that it is objectionable to you and that you would like it stopped.
2. If the behavior does not stop, report the offender to your immediate supervisor or to his or her supervisor if your supervisor is the offender. Some restaurants have a telephone hotline for this purpose.
3. If using the internal complaint procedure does not correct the problem, file a complaint with your state department that deals with human rights or with the Equal Employment Opportunity Commission (EEOC).

The management of most restaurants has a zero-tolerance policy that states they will not tolerate any acts of harassment, intimidation, or threats among their employees. If reported, most offenders will be investigated discreetly and fairly. Management should take appropriate action, whether it is against innocent injury or criminal intent to harm.

KEEPING VIOLENCE AND DRUGS OUT OF THE WORKPLACE

Violence is rough or harmful action or treatment. Everyone shares responsibility for maintaining a safe work environment. When someone acts out of the ordinary, management should be advised immediately. Unreported situations can have potentially violent consequences.

A potentially violent individual is frequently someone who is depressed, is a loner who intimidates those around him or her, or is lacking self-worth. It can often be a person who is a constant complainer or has a history of violence. Violence in the workplace may be a direct result of problems in the home. Drug and/or alcohol abuse often play a role in violent incidents.

It is against the law to use controlled substances. It is crucial for all employees to be drug-free at the workplace. An employee who uses illegal drugs is more likely to be involved in accidents on the job, typically needs more sick leave, and is more frequently late for work than other employees. Overall productivity also tends to deteriorate, and team goals become less of a priority.

SAFETY AND PREVENTING ACCIDENTS

Safety is freedom from harm or danger. It is important for employees and management to work together to maintain a safe and secure restaurant environment. Servers should be alert to any hazards they encounter. Most accidents are avoidable if a problem is noticed and solved in time. Addressing unsafe work routines, blocked exits, chipped or broken serviceware, and grease and food spills immediately upon notice will prevent accidents from occurring (see Chapter 6, Safety, Sanitation, and Emergency Procedures).

TAKING SANITATION SERIOUSLY

The serving team has almost as much contact with foods served to guests as the kitchen staff. **Sanitation** is the process of working out ways to improve health conditions. It is of the utmost importance that servers adhere to strict sanitation guidelines when handling food to avoid spreading diseases that may be distressing or life-threatening to guests. Good sanitation practices include washing hands and handling serviceware properly. Servers must be very conscientious in this regard. (Sanitation guidelines and proper food-handling techniques are covered in depth in Chapter 6.)

KEY TERMS

Servers	Grooming	Expeditor
Waitstaff	Organization chart	Teamwork
Service	Busser	Diversity
Hospitality	Head server	Harassment
Suggestive selling	Host	Violence
Uniform	Maître d'hôtel	Safety
Hygiene	Dining room manager	Sanitation

REVIEW

1. Explain the difference between service and hospitality. Give two examples of service and two examples of hospitality.
2. What advantages of a serving position can you list other than those that are mentioned in this chapter? What disadvantages can you think of?
3. What qualifications for a serving position can you identify other than those mentioned in this chapter?
4. List all the hygiene and personal grooming practices that should be given attention in order to promote cleanliness, good health, and neat appearance on the job.
5. Why should a server be knowledgeable about the organization of employees in a restaurant?
6. What part does teamwork play in the operation of a restaurant?
7. Why is it important to accept the diversity of all coworkers and guests?
8. What is the procedure for handling sexual harassment?
9. What is one way that you can reduce the possibility of a violent situation in the workplace?

PROJECTS

1. Design a server's self-evaluation sheet. List both the ideal personality and appearance qualifications for a serving position on the left side of the paper; think of other qualifications you can add to those discussed in this chapter. Across the top, write a rating scale: Poor, Fair, Good, and Excellent. Rate yourself by checking the appropriate column for each qualification. Set a goal to improve yourself in all areas not marked Good or Excellent.
2. Observe the servers in a restaurant of your choice, and note whether their appearance is satisfactory or unsatisfactory.

3. Draw an organization chart of a restaurant of your choice. Indicate who would be your immediate supervisor if you were a server there, and why.
4. Have a group discussion and decide the best course of action to handle the following situations:
   a. The chef has made a mistake on your order.
   b. You have taken the order, and the guest states that you did not get the order correct.
   c. You present the check to your table of guests, and they ask you to divide the check for them because they want to pay separately.
   d. You are an above-average server and think that the policy of sharing tips equally with fellow workers is unfair.
   e. You have come down with a bad cold the night before you have to go to work.
5. Have a group discussion about the best plan of action to implement for the following situation: A new employee has just begun to work at the restaurant. She is from another country and speaks with an accent. She needs to be accepted onto the team. What can you, as a peer server, do to help her overcome any anxiety she may have?

CASE PROBLEM

Handling Harassment

While you are working as a server, you are approached by your supervisor, who makes remarks with sexual overtones. You feel uncomfortable about these remarks. The state in which you are working has strong sexual harassment laws, and you are aware of them. You want to curtail this situation before it leads to something you do not want to happen. Answer the following questions:

- What are your rights?
- What steps can you take to make sure your supervisor understands your feelings and that allow you to continue your job in that establishment?
- To whom can you report this situation if it goes beyond your control?

Chapter Two

Types of Establishments, Types of Service, and Table Settings

Objectives

After reading this chapter, you will be able to:

- ✔ Differentiate between many types of establishments.
- ✔ Identify the characteristics of French service, Russian service, English service, and American service.
- ✔ Describe the arrangement of serviceware for the American breakfast, lunch, and dinner cover.
- ✔ Explain where food items are placed in the American cover as they are served.
- ✔ Define banquet service, family-style service, buffets, salad bars, oyster bars, and dessert tables.

Every restaurant has a unique identity depending on the amenities offered to guests. This chapter discusses how restaurants are classified depending on their amenities and mode of operation. Also discussed in this chapter are several different types of service used in restaurants. Most types of service originated in the private homes of European nobility, and over the years have been modified for restaurant use. Today, each type retains particular distinguishing features, although some restaurants have combined features of two or more serving styles to accommodate their menu, facilities, and mode of operation. The four traditional types of service discussed in this chapter are French, Russian, English, and American, and the **cover**, the arrangement of china, silverware, napkin, and glassware at each place setting, is diagrammed. Other popular types of service explained in this chapter are the banquet, family-style, buffets, salad bars, oyster bars, and dessert tables.

Types of Establishments

Many terms are used to describe types of restaurants, such as tearoom, family-style, upscale, casual, theme, and quick-service. The fact of the matter is that restaurants are subjectively classified by the style of operation. There are no clear characteristics for each type of establishment. Traits of one style of restaurant overlap with traits of another, creating a unique identity. For our purposes here, we could say that types of establishments form somewhat of a continuum. On one end of the continuum would be restaurants with minimum service, no covers or simple placemats, a quick pace, fast foods prepared easily, and a general informality to the entire operation. The other end of the continuum would be the establishments with luxurious surroundings; table linens; silver serviceware; china; crystal glassware; flowers; soft music; an unhurried pace; skilled servers, and expensive, well-prepared, and well-presented foods and wines. Family-style restaurants, diners, and some chain restaurants would be on one end of the continuum; trattoria, bistros, and supper clubs in the middle; and classic gourmet, upscale, and fine-dining restaurants on the other. No matter which combination of amenities make up the whole, the guest has every right to expect a smile, that his or her order be correctly taken and delivered, and that the check is presented promptly and for the correct amount regardless of the setting. Good service is still the key to running any successful operation.

Types of Table Service

FRENCH SERVICE

French service is a formal type of service originated for European nobility and currently enjoyed by the few who can afford the time and expense of meals served

FIGURE 2-1 French Service, Tableside Cooking: In French service, food is cooked in front of the guests on a small spirit stove. Photo by S. Dahmer

in this manner. This type of service is used in upscale restaurants, elegant hotel dining rooms, cruise ships, resorts, and casinos.

In French service, the food is either cooked or completed at a side table in front of the guests (Figure 2-1). The food is brought from the kitchen to the dining room on heavy silver platters and placed on a cart called a **guéridon.** A small spirit stove called a **réchaud** is used to keep the food warm. The food is completed by cooking, deboning, slicing, and garnishing as necessary and served to the guests on heated plates. Only those foods that can be cooked, assembled, or completed in a reasonably short time are prepared in front of the guests. Typical specialties that may be served in the French style are La Salade César (Caesar salad), Le Tournedos au Poivre (pepper steak), and Les Crêpes Suzettes (crêpes in orange sauce).

French service employs two servers working together to serve the meal and may include a captain to seat guests and a wine steward to serve wine. The principal server is the **chef de rang** (or experienced server), who seats the guests when a captain is not present, takes the order, serves the drinks, prepares some of the food with flourish at the guests' table, and presents the check for payment. The assistant is the **commis de rang**, who takes the order from the chef de rang to the kitchen, picks up the food and carries it to the dining room, serves the plates as dished up by the chef de rang, clears the dishes, and stands ready to assist whenever necessary. All food is served and cleared from the right of the guests except for butter, bread, and salad, which should be placed to the left side of the guests.

Finger bowls—bowls of warm water with rose petals or lemon slices in them—are served with all finger foods and at the end of the meal. The finger bowl is set on an **underliner**, a small plate with a doily, and placed, with a clean napkin, in front of the guests. Soiled dishes are cleared only when all guests have completed their meals.

French Table Setting

The French cover includes an hors d'oeuvre plate (or show plate), napkin, dinner fork, dinner knife, soup spoon, butter plate, butter spreader, dessert fork and spoon, and a water or wine glass. The French arrangement of serviceware is shown in Figure 2-2.

Advantages and Disadvantages of French Service

The advantages of French service are that guests receive a great deal of attention, and the service is extremely elegant. The disadvantages are that fewer guests may

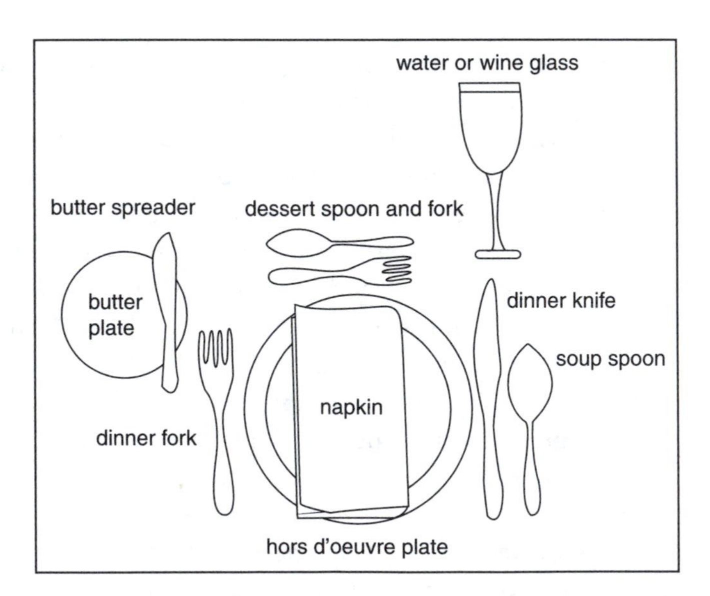

FIGURE 2-2 French Service Cover: Serviceware in French service includes a butter plate, butter spreader, hors d'oeuvre plate, napkin, dinner fork, dinner knife, soup spoon, dessert fork and spoon, and water or wine glass.

be served, more space is necessary for service, many highly professional servers are required, and service is time-consuming.

RUSSIAN SERVICE

Russian service is similar to French service in many respects. It is very formal and elegant, and the guest is given considerable personal attention. It employs the use of heavy silver serviceware, and the table setting is identical to the French setup. The two major differences are that only one server is needed and that food is fully prepared and attractively arranged on silver platters in the kitchen.

To serve, the server places a heated plate before each guest from the right side, going around the table clockwise. Then the server brings the platters of food to the dining room from the kitchen and presents them to the guests at the table.

Standing to the left of each guest and holding the platter of food in the left hand, the server shows each guest the food and then, using a large spoon and fork in the right hand, dishes up the desired portion on the guest's plate (Figure 2-3). The server continues serving counterclockwise around the table and then returns the remaining food to the kitchen. As in French service, finger bowls and napkins are

FIGURE 2-3 Russian Service, Using Serving Spoon and Fork: When serving food from a pan or platter, the server uses a serving spoon and fork in one hand to place food on the guest's plate. Photo by S. Dahmer

served with the meal, and soiled dishes are cleared when all guests have completed the meal.

Advantages and Disadvantages of Russian Service

The advantages of Russian service are that only one server is needed and that this service is as elegant as French service, yet faster and less expensive. No extra space is needed for special equipment, such as the guéridon.

The disadvantages of Russian service are the large investment in silver serviceware and the number of platters needed, especially when every guest in a party orders a different selection. For this reason, Russian service is particularly useful at banquets where every guest receives the same food selection. Another disadvantage is that the last guest served at the table must be served from the less well-displayed food remaining.

ENGLISH SERVICE

English service is used occasionally for a special dinner served in a private dining room of a restaurant, but it is more typical of a meal served by servants in a private home.

The food on platters and the heated plates are brought from the kitchen and placed before the host at the head of the table. The host or one of the servers carves the meat, if necessary, and dishes up the entrée and vegetable on individual plates. He or she hands the plates to the server standing to the left, who serves the guest of honor and all other guests. Dessert may also be served in this manner. All sauces and side dishes and, in some cases, vegetables are placed on the table to be passed by the guests.

Advantages and Disadvantages of English Service

The advantage of English service is that it involves a great deal of showmanship for a special occasion. The disadvantages are that the host may be required to do a lot of the work by dishing up some of the food, and with only one person serving the meal, the service can be very time-consuming.

AMERICAN SERVICE

American service is less formal than French, Russian, or English service. It is the most prevalent style of service in restaurants in the United States. In American service, food is dished up on plates in the kitchen. Except for the salad and the bread and butter, most of the food is placed on the dinner plate. Usually, only one server presents the meal. Food is served from the left of the guest, beverages are served from the right, and soiled dishes are cleared from the right.

The American breakfast and lunch table setting differs from the American dinner setting. Breakfast and lunch are simple meals and require only a limited amount of serviceware. Dinner involves more courses and more serviceware.

American service can be simple and casual or complex and elegant. It can be used to serve the guest who wants a quick, filling meal at a casual restaurant with simple service. For instance, it can be used at a counter, diner, or family-style restaurant, where casual tableware and placemats are the norm, and where self-serve bars such as salad bars are common. American service can also be used to serve the guest who intends to be entertained for the evening at a five-star establishment. It can be used to present food elegantly in distinguished gourmet restaurants with formal table settings and the use of complex serving skills and showmanship. The rest of this book explains American service in detail.

American Breakfast and Lunch Cover

Serviceware for the American breakfast and lunch cover includes a dinner fork, dinner knife, teaspoon, napkin, bread and butter plate, butter spreader (optional), and water glass.

The breakfast or lunch cover is placed approximately one inch from the edge of the table. The napkin is in the center of the cover. To the left of the napkin is the fork, and to the right of the napkin is the knife, with the blade facing toward the napkin. The teaspoon is to the right of the knife. The water glass is placed above the tip of the knife. The bread and butter plate is placed above the tines of the fork. Traditionally, a small butter spreader was placed on the rim of the bread and butter plate, but today only a few restaurants use it. Figure 2-4 shows the initial American breakfast and lunch cover.

When coffee is served, the cup and saucer are placed to the right of the teaspoon. Breakfast toast or a luncheon salad is placed to the left of the fork. Food on the dinner plate is placed directly in the center of the cover after the guest has removed the napkin. Side dishes and accompaniments are placed in a convenient location on the table when served. Figure 2-5 shows the place setting, including the dishes served during the meal.

American Dinner Cover

Serviceware for the American dinner cover includes two dinner forks, dinner knife, butter spreader, two teaspoons, service plate (optional), napkin, bread and butter

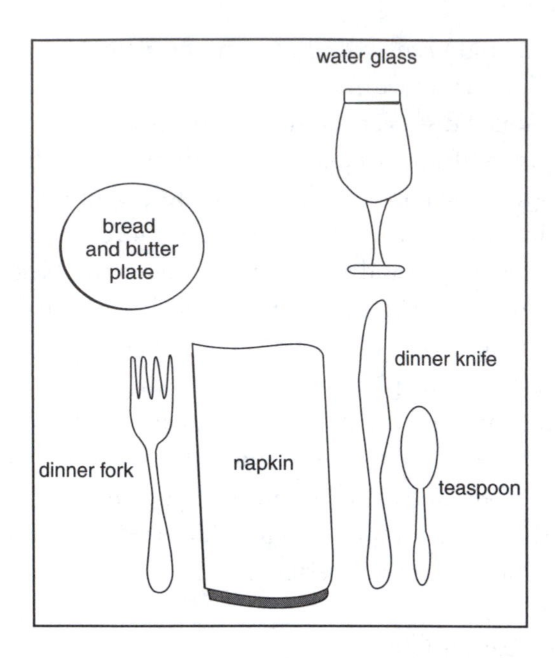

FIGURE 2-4 Initial American Breakfast and Lunch Cover: The initial American breakfast and lunch cover is set with a napkin, dinner fork, dinner knife, teaspoon, bread and butter plate, and water glass. An optional butter spreader may be placed on the bread and butter plate.

plate, and water glass. Today, the butter spreader and second teaspoon are often eliminated, but traditionally and for this discussion, they will be included.

The dinner cover is placed approximately one inch from the edge of the table. The napkin is placed on a service plate or by itself in the center of the cover. The two dinner forks are to the left of the napkin. The dinner knife is to the immediate

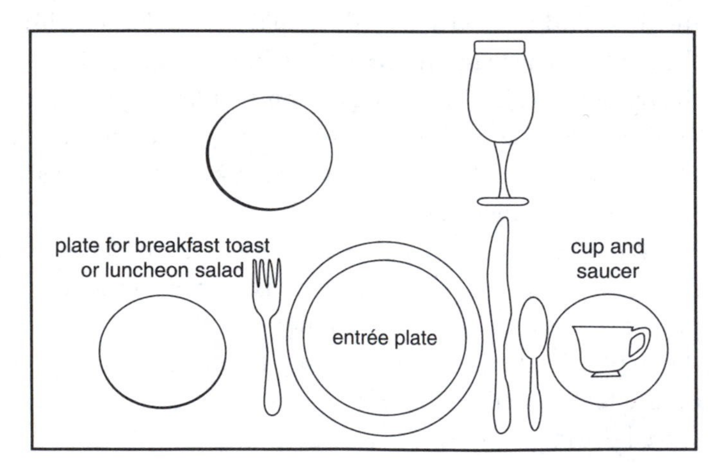

FIGURE 2-5 American Breakfast and Lunch Cover as Food Is Served: As food is served for the American breakfast or lunch, the dishes are placed in specific locations of the cover as diagrammed. The breakfast toast or luncheon salad is placed to the left of the fork, the entrée is centered, and the cup with saucer is placed to the right of the spoon.

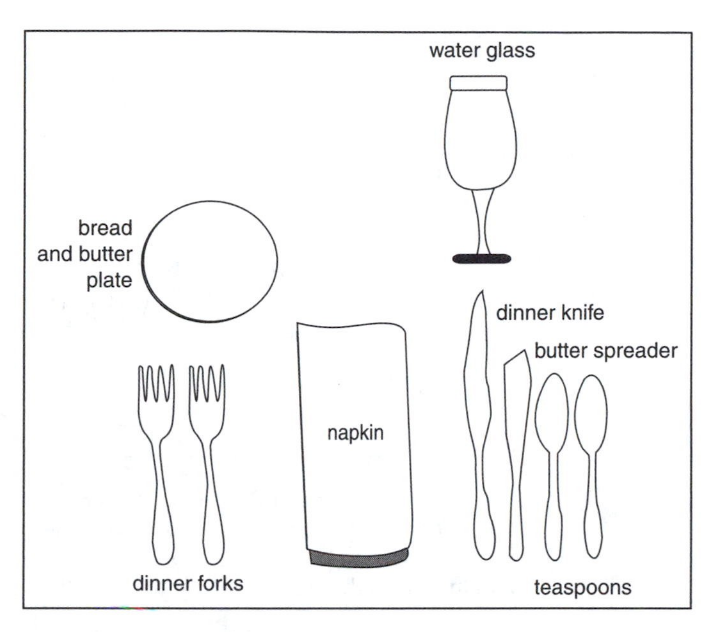

FIGURE 2-6 Initial American Dinner Cover: The initial American dinner cover includes a napkin, two dinner forks, a dinner knife, butter spreader, two teaspoons, bread and butter plate, and water glass. An alternative cover is to have the butter spreader on the bread and butter plate.

right of the napkin and then, in order, is the butter spreader and two teaspoons. The blades of the knives face the napkin. The water glass is placed directly above the knives. The bread and butter plate is centered above the forks. An alternative American cover is to have the butter spreader placed on the bread and butter plate. The initial American dinner cover is shown in Figure 2-6.

When soup or an appetizer is ordered, it is placed on an underliner and served in the center of the cover. The salad is placed to the left of the forks, and coffee, when served, is placed to the right of the spoon. The food on the dinner plate is placed in the center of the cover. Special-purpose silverware, such as a soup spoon with soup or a steak knife with steak, is brought in as needed. Rolls, accompaniments, and side dishes are placed in convenient locations on the table. Diagrammed in Figure 2-7 is the place setting, with the dishes served during the course of the meal.

BANQUET SERVICE

Banquet service involves serving a meal to a group of people who are celebrating, gathering for a special occasion, or honoring special guests. The menu, number of guests, and time of service are predetermined, and the banquet is well organized in

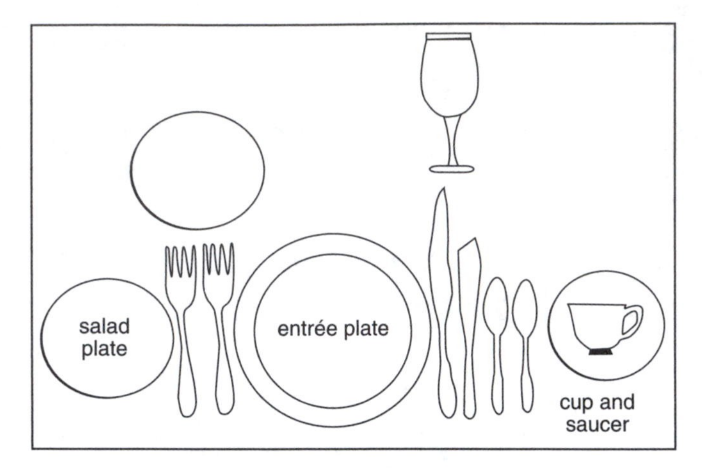

FIGURE 2-7 American Dinner Cover as Food Is Served: As the dinner is served using the American dinner cover, the salad is placed to the left of the forks, the entrée is centered, and the cup with saucer is placed to the right of the spoons.

advance (Figure 2-8). Banquet service is offered in hotels, resorts, country clubs, casinos, and restaurants that have conference rooms for holding meetings.

The server generally sets the tables with American settings modified according to the particular menu. For example, soup spoons or steak knives may be part of the initial cover if soup or steak is to be served (Figure 2-9). Occasionally, French, Russian, or buffet service is used at a banquet, and the table is set accordingly.

If a cold course is planned, such as a salad, it is placed on the table just before the guests are seated. Also at this time, ice water is poured, butter pats are placed on the bread and butter plates, and baskets of hot rolls are arranged on the tables.

The food is put on plates in the kitchen and served to the guests in the usual American serving style (see Chapter 5, Serving the Meal) or in French, Russian, or buffet style, as predetermined. The head table is served first, then the rest of the tables. Water and coffee are replenished periodically.

If the guests will remain seated for entertainment or a speaker after dinner, be sure the tables are cleared and tidy at the completion of the meal. Because the banquet is paid for in advance, the server does not need to present a check or collect payment. The tip is often included in the contract made by the catering manager and the client arranging the banquet.

Advantages and Disadvantages of Banquet Service

The advantage of banquet service is that the menu and serving time are predetermined, which makes service a simple routine, accomplished by fewer servers than

FIGURE 2-8 Banquet: The menu, number of guests, and time of service of the banquet are predetermined in advance by a contract between the client and the banquet manager. Photo by S. Dahmer

FIGURE 2-9 Banquet Cover: For a banquet, the server generally sets the tables with American settings modified according to the particular menu being served.
Courtesy of Radisson Hotel South and Plaza Tower, Minneapolis, Minnesota; Photo by S. Dahmer

needed for other types of serving. A disadvantage of banquet service is that guests receive very little personal attention. They are usually seated in close quarters, which make proper service difficult.

FAMILY-STYLE SERVICE

Family-style service is a modification of American service and somewhat more informal. All necessary preparation, such as cooking foods and slicing meats, is done in the kitchen. The food is then placed in large bowls and on platters and is properly garnished. The server serves the bowls and platters by placing them in the center of the table. The food is passed around the table by the guests, who help themselves to the portions desired.

The amount of service required of the server is minimized, because the server simply sets the food on the table, pours the beverages, brings the appropriate condiments, and clears away the soiled dishes. Initially, the platters and bowls contain only enough servings for the number of guests in the party. If the family-style service has an all-you-can-eat feature, servers must refill serving containers when requested. Usually, the American cover or a modification is used. Serving utensils are brought with the platters and bowls of food.

Advantages and Disadvantages of Family-Style Service

This simplified manner of service is advantageous to new waitstaff who have not learned the proper details of serving. It is fast because the guests actually serve themselves; servers can serve more people than when a more formal type of service is used.

The disadvantages are that guests receive less personal attention and must serve themselves from a food platter that becomes less attractive as other guests serve themselves.

BUFFETS

With **buffet service**, guests select their meals from an attractive arrangement of food on long serving tables (Figure 2-10). The guests either help themselves or are served by chefs standing behind the buffet tables. The service usually combines both methods—the guests select relishes, salads, and vegetables themselves, and the meat is carved and served to the guests by chefs. Silverware and napkins may be conveniently located on the buffet table for the guests to pick up with their meals,

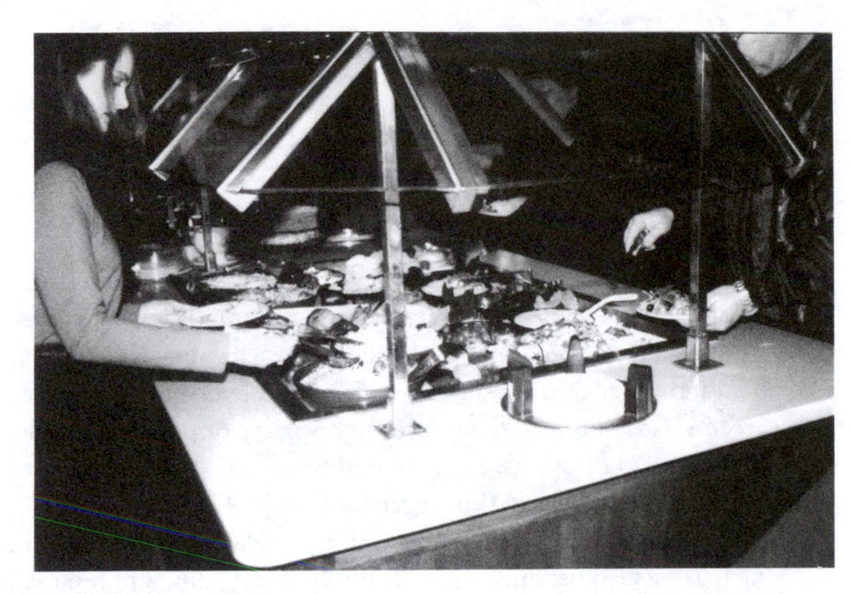

FIGURE 2-10 Buffet: In buffet service, guests help themselves from an attractive arrangement of food on serving tables.
Courtesy of Radisson Hotel South and Plaza Tower, Minneapolis, Minnesota; Photo by S. Dahmer

or a complete cover (usually American), including rolls, butter, and condiments, may be preset at the dining tables.

A **smorgasbord** is a buffet featuring a large selection of food with Scandinavian selections, such as cheese and herring. In some places, it is a set-price, self-service buffet of any kind of food. Usually, guests may come back to the smorgasbord table and get more food as often as they desire.

The job of the server varies, depending on the design of the buffet. The servers may serve only beverages and dessert, or they may serve several courses, such as the appetizer and soup, at the guests' tables. To maintain the sanitary condition of the buffet foods, remove soiled tableware and notify guests that they are to use clean plates each time they return to the buffet.

Advantages and Disadvantages of Buffet Service

One advantage to buffet service is that food can be displayed in a very attractive manner. However, this can quickly become a disadvantage if care is not taken to keep the food selections fresh and complete. Another advantage is that servers can attend to many guests at one time, but guests receive less personal attention than with table service.

SALAD BARS, OYSTER BARS, AND DESSERT TABLES

In three other variations of buffets, only the salad, seafood, or dessert is served buffet style. The rest of the meal is served in the usual manner.

Salad Bars

A **salad bar** is a self-service concept in which each guest is given the opportunity to prepare his or her own salad from an attractive array of fresh vegetables and fruits that have been cleaned and sliced or quartered (Figure 2-11). Bowls and salad plates are available at one end of the salad bar. Guests prepare their own tossed green salads and help themselves to a variety of prepared salad accompaniments, such as salad dressings, crackers, and bread sticks. More elaborate salad bars offer soups, pasta salads, cold cut vegetables, fruits, hard-boiled eggs, grated cheese, and crouton and seed toppings. And even more elaborate salad bars may include dishes such as pickled herring, sardines, thinly sliced ham, and tuna salad. Salad bars are available in casual, informal restaurants.

FIGURE 2-11 Salad Bar: A salad bar is a self-serve concept in which each guest prepares his or her own salad from a selection of greens, fruits, vegetables, and accompaniments. Some salad bars have an extensive selection of prepared salads, soups, cold meats, and cheeses as well.
Photo by S. Dahmer

A clear, protective panel, called a **sneeze guard** or food guard, is mounted above the salad bar to keep the ingredients sanitary. Guests must also use a clean plate or bowl for each return trip. The sneeze guard and clean plate for refills ensure that salads are protected from contamination.

The duties of the servers are first to take the guests' meal and wine orders and then to inform them when and how to begin the salad bar. Remind guests that they must use a clean plate each time they go to the salad bar for refills. Servers also assist the kitchen staff in the upkeep of the salad selections by informing them when food needs replenishing. Servers should remove soiled salad dishes from tables as they accumulate and keep beverages fresh during this course.

Oyster Bars

An **oyster bar** is a buffet featuring oysters on the half shell and various seafood and mustard sauces that complement the oysters. Sometimes boiled shrimp or other appetizer seafood extends the selection. The chef may want to display these foods in a nautical setting of shells, ice chips, nets, and diving relics to add to their appeal.

Dessert Tables

A tantalizing display of tortes, pies, cakes, cream puffs, éclairs, fresh fruits, and soft cheeses displayed in buffet fashion constitutes a **dessert table**. Dessert plates and forks are at hand at the dessert table. Desserts are usually precut into portions, and guests help themselves.

Well-displayed examples of desserts can also be brought to the guests' table on a tray (Figure 2-12) or on a dessert cart with wheels. Servers bring the tray or cart at the end of the meal and sell from the appetizing selection at hand.

Advantages and Disadvantages of Salad Bar, Oyster Bar, and Dessert Table Services

As with conventional buffet service, the food at the salad and oyster bars and dessert table can be arranged very attractively. Less work is involved for the server because the guests take care of getting their own soup, salad, bread, appetizer, and dessert. The server has more time to serve many guests, which increases opportunities for extra tips.

The disadvantage of salad or oyster bars or dessert tables is that, like a buffet, they must be replenished continually to look attractive. In addition, the timing of

FIGURE 2-12 Dessert Tray: A selection of delectable desserts can be brought to the guests' table on a tray. Courtesy of PhotoDisc, Inc.

the meal can be upset; the server must be able to coordinate the self-service of the guest at the salad or oyster bar with the rest of the meal.

Dessert carts or trays, unlike self-serve bars and tables, require more time on the part of the server. The server must push or carry them to the table and then sell and serve the desserts.

KEY TERMS

Cover
French service
Guéridon
Réchaud
Chef de rang
Commis de rang
Finger bowls
Underliner
Russian service
English service
American service
Banquet service
Family-style service
Buffet service
Smorgasbord
Salad bar
Sneeze guard
Oyster bar
Dessert table

REVIEW

1. Define cover and underliner.
2. Describe in your own words how service might differ in a very casual, informal restaurant from service in a high-end, upscale restaurant.
3. What type of restaurant would offer the following:
   a. smorgasbord
   b. spumoni
   c. thick steaks
   d. wine service
   e. barbeque
   f. sauerbraten
   g. gourmet teas
4. Why do very few restaurants use French service?
5. How are French service and Russian service alike? How do they differ?
6. When is English service used today?
7. Why is American service used in most restaurants today?
8. Why is American service sometimes called plate service?
9. Why do the amount and arrangement of serviceware in a cover vary? What other serviceware arrangements have you seen?
10. How is a banquet set up prior to the seating of guests?
11. What is the distinguishing feature of family-style service?
12. How are buffets, salad bars, oyster bars, and dessert tables similar? How do they differ from one another?
13. From your experience, give examples of how two types of service are used together.

PROJECTS

1. Make two lists under the heading "Types of Restaurant Establishments." Title one list "Simple, Informal Restaurants" and title the other "Upscale, High-End Restaurants." Under each heading, list as many characteristics of that type of restaurant operation as you can. When you are finished, discuss the many types of restaurants that fall between these two extremes and the features they use from each list to create their own identity.
2. Make a chart for the eight types of service discussed in this chapter. List the types of service down the left margin. Across the top of the chart, label the vertical columns with the following headings: Distinguishing Features, Server's Responsibilities, Advantages, and Disadvantages. Complete the chart by using information from this chapter.
3. Using proper serviceware, set up the American breakfast and lunch cover and the American dinner cover. In a training session, point out similarities and differences between the settings. Identify the purpose of each serviceware piece, and show the placement of various foods as they are served.
4. Observe different types of service in various restaurants, and discuss with other trainees what you have learned.

CASE PROBLEM

Buffet Problem

As a server, you observe children going to the buffet and dipping the serving utensil from one food item into another. Also, you notice one child picking up food with his hand. Other guests observe the children. The parents of the children appear to be unaware of where their children are and what they are doing. Answer the following questions:

- How would you approach the children's parents concerning this problem?
- What would you do with the food that was contaminated?
- What steps should be taken to avoid an occurrence of this sort of situation in the future?

Chapter Three

Before the Guests Arrive

Objectives

After reading this chapter, you will be able to:

- ✔ Explain the concept of a station.
- ✔ Understand why it is important for a restaurant to take and honor reservations.
- ✔ Describe how to prepare the tables and tablecloths.
- ✔ Fold several styles of decorative napkins.
- ✔ List many supplies that should be stocked at the sidestand.
- ✔ Differentiate between many types of menus and the categories in each.
- ✔ Define food preparation methods.
- ✔ Identify many meal accompaniments that complement food items.
- ✔ List the closing duties needing to be done in preparation for business the next day.

Servers have many responsibilities to attend to prior to serving guests in a restaurant. Servers are first assigned the area of the restaurant in which they will work and the tables they will serve, holding some of these tables for reservations. They must then attend to sidework. **Sidework**, also called **mise en place**, is a term designating all of the duties the server performs other than those directly related to serving the guests. Sidework includes the opening duties, such as preparing the dining room and studying the menu, as well as closing duties, leaving the work area in proper order upon completion of the shift.

Station Assignments

A **station** is a section of the dining room assigned as a work area to a server. Each station has seating for about a dozen or more guests at tables, booths, or counters. Ideally, a dining room should be divided into stations that are equal to one another in the number of people they seat, in their distance from sidestands and the kitchen, and in desirability of seats to the guests. Of course, this balance is impossible in most dining rooms, because there will always be less desirable seats near kitchen and washroom entrances and away from scenic views.

Because stations are not equally desirable from a seating and serving standpoint, dining room managers often assign stations to servers on a rotational basis—servers take turns from day to day serving in the best stations.

In some restaurants, servers with seniority have permanent stations that are larger or more desirable than others. This assignment is made because these servers are experienced and can handle more guests and because certain guests request a particular server and seat. A new server may be assigned a less desirable station, which provides an opportunity to gain experience with a smaller number of guests.

For convenience, tables are often numbered, and stations are assigned by giving the numbers of the tables to a server (Figure 3-1). The server then uses these numbers on orders and guest checks to identify the party of guests being served.

Reservations

Generally, one member of the restaurant staff, a **receptionist** or **reservationist**, will handle reservations. **Taking reservations** means promising a table to guests who call or e-mail in advance. For instance, a guest may call on Friday to reserve a table for four for dinner at 8:00 P.M. on Saturday evening. The reservationist will record a name, number of guests, a phone number, or even a credit card number to hold the reservation, and assign the table to a station. Special requests such as server-of-choice, table-of-choice, birthday cake, or high chair should be honored, if possible. The reservationist may want to confirm the reservation with a phone

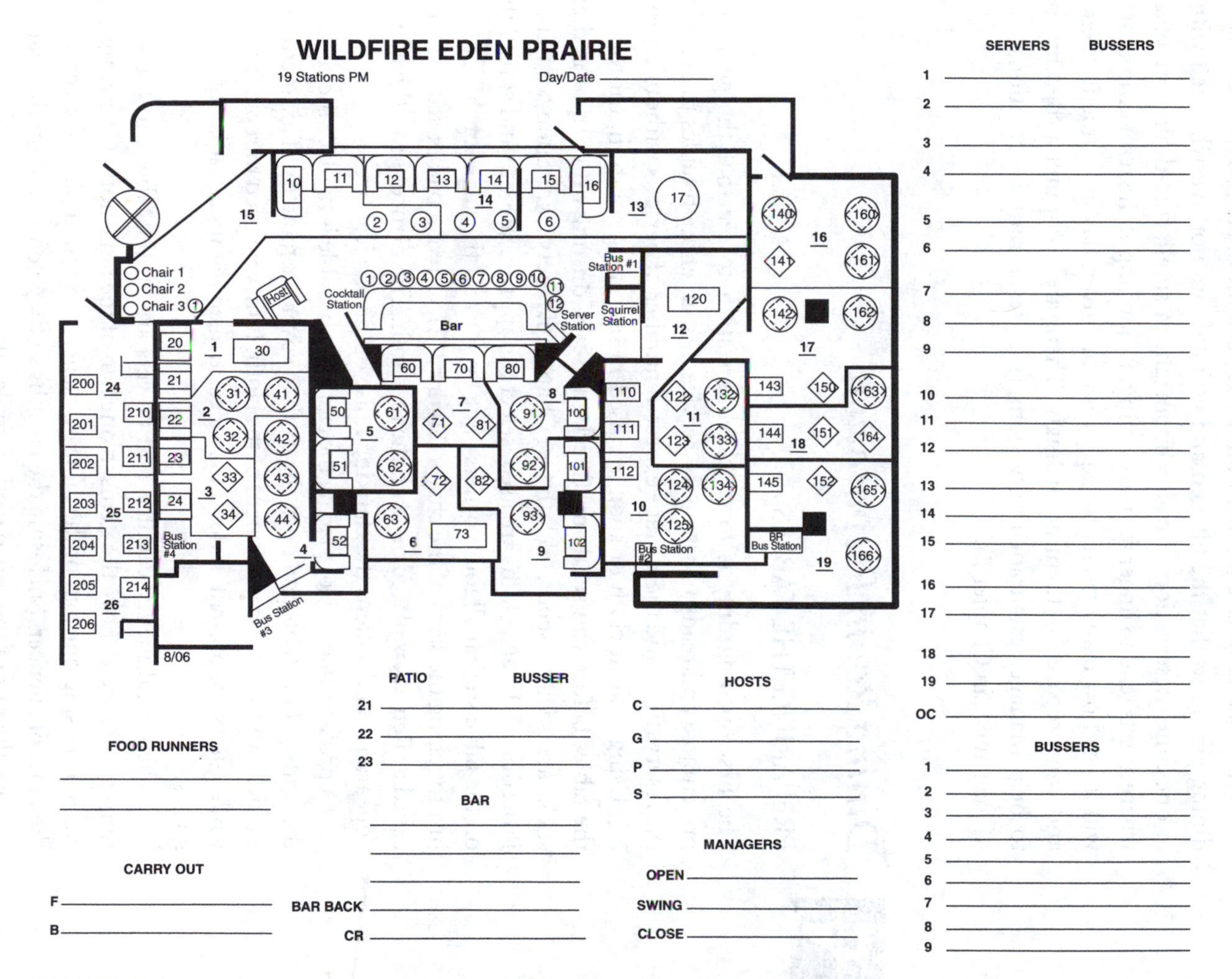

FIGURE 3-1 Station Assignments: Tables in dining rooms are often numbered, and stations are assigned by giving the numbers of the tables to the server. Courtesy of Lettuce Entertain You Enterprises, Inc.

call just prior to the date. Guests will then be expected at an appointed time and may be seated immediately upon arrival.

The reservationist's job is to fill the restaurant at staggered intervals over the dining period so that the kitchen and waitstaff are not overwhelmed with guests at one time. Reservations are also important in small, fine-dining establishments where guests tend to linger for the evening. By looking at the reservation list, servers will know how many guests to expect on any given day and time in their station and can prepare for them accordingly. There are computer programs available to help manage restaurant reservations (see Restaurant Reservations and Table Management in Chapter 7).

Dining Room Preparation

PREPARING THE TABLES

The first opening duty is to check your station to see that the general area is presentable and ready to set up for service. Set up enough tables to accommodate the reservations and the average expected number of persons without reservations.

Using a clean cloth or sponge rinsed in sanitizing solution, thoroughly wash the tables before you set them. Check the seats, dusting off crumbs and cleaning sticky areas with a separate cloth (see Chapter 6, Safety, Sanitation, and Emergency Procedures). If tablecloths are used, select the appropriate size and spread the cloth on the table so that all four corners hang evenly and the edges of the tablecloth are just touching the seats of the chairs (Figure 3-2). Often, a pad or second tablecloth, called a **silencer**, is placed beneath the top cloth. The silencer gives the table a better appearance and softens the clanking noise of the serviceware.

A professional way of placing the cloth on the table is to place the center fold on the center of the table and open the cloth to cover the tabletop. This method assures a quick, well-centered placement of the cloth. It may be used to replace soiled cloths while guests are present, if done so in a controlled manner that is not distracting to guests.

When condiments, candles, and flowers are on the table and the soiled cloth must be changed, move the items to one half of the tablecloth. Gather up the soiled cloth, exposing one half of the table or silencer, and then place the center items on the table or silencer. Enclosing the crumbs so they do not fall on the seats or floor, remove the rest of the soiled cloth completely.

Replacing the cloth is the reverse operation. With center items remaining at the edge of the table, place the center fold of the tablecloth at the center of the table.

FIGURE 3-2 Tablecloth Placement: The proper way of arranging the tablecloth is to have the edges of the cloth just touch the chairs. Guests approaching the table get a favorable impression of the meal to come as they observe the even arrangement of the cloth, napkins, and place settings.
Photo by S. Dahmer

Fold up the top half so the center items may be placed on the surface of the cloth. Then open the cloth completely and arrange the condiments, centerpieces, and other items (Figure 3-3). If placemats are used, arrange them neatly on the clean tables. Some restaurants do not use a tablecloth or placemats. In any case, the table needs to be clean.

After the tablecloths or placemats are arranged properly, set up the covers. A **cover** consists of china, silverware, napkins, and glassware at each place setting. The

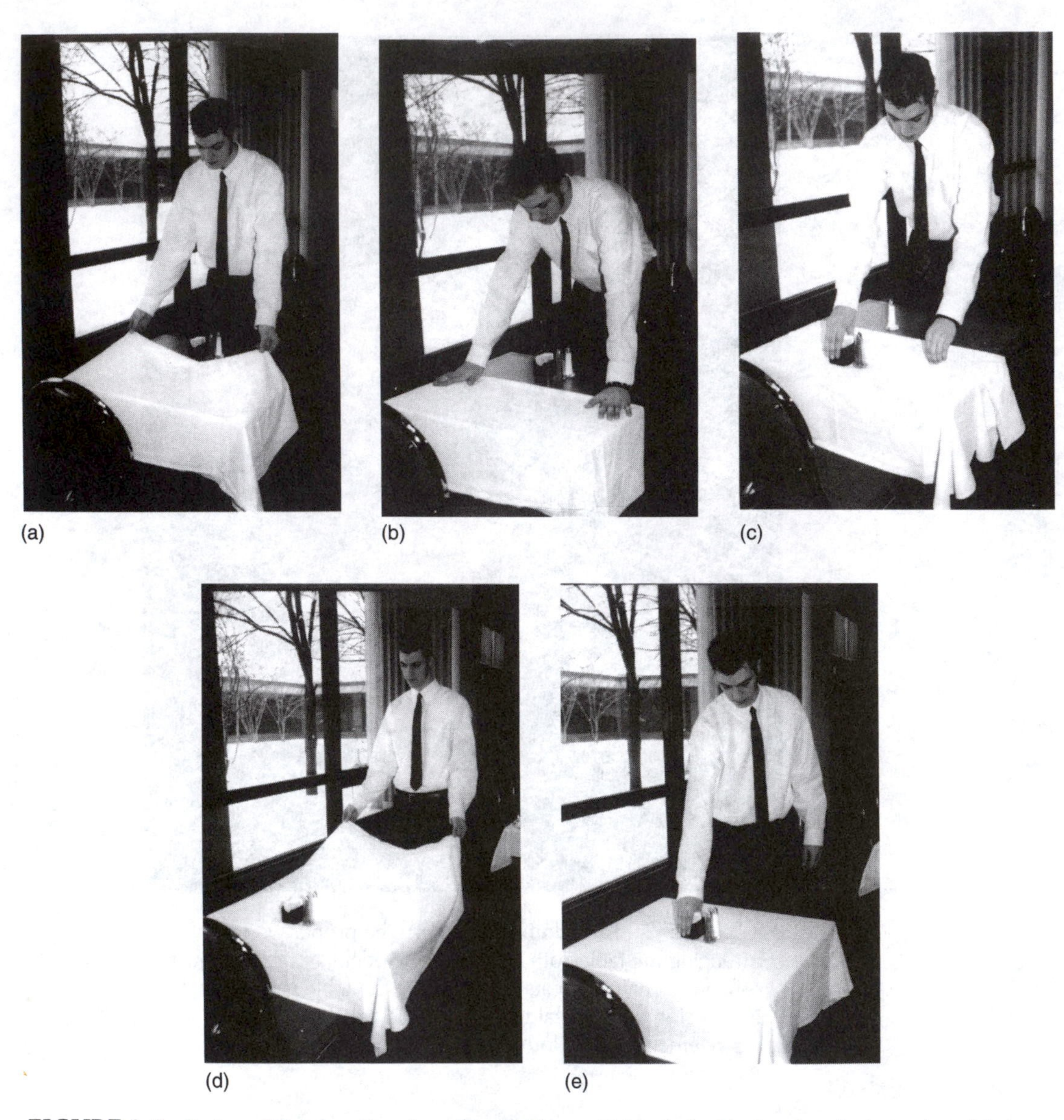

FIGURE 3-3 Series of Pictures Showing How to Place a Tablecloth: Placing the cloth on the table is illustrated: (a) with the condiments or vase moved to the edge, place the center fold of the cloth on the table; (b) position the cloth so it drapes evenly; (c) open the cloth, gently gather or fold it up at the center of the table, and transfer the condiments or vase to the half already spread; (d) spread the cloth over the rest of the table; and (e) center the condiments or vase.
Courtesy of Hennepin Technical College, Eden Prairie, Minnesota; Photos by S. Dahmer

amount of serviceware and the arrangement depend on the type of service and the meal to be served (see Chapter 2, Types of Establishments, Types of Service, and Table Settings).

Carry supplies of chinaware, glassware, silverware, and napkins to the table on clean trays. Handle china by the edges, glassware by the bases or stems, and silverware by the handles as you set up the covers (see Figure 6-5, Handling Food and Tableware). Set aside any serviceware that is soiled, and return it to the kitchen. Discard any chipped or cracked glassware or china.

After the covers have been set, check to see that all centerpieces are fresh and clean and that candles are replaced or lights are in working order. When **table tent menus**—small menus designed to stand vertically—are used, place them uniformly on all tables.

FOLDING THE NAPKINS

The fold of a cloth napkin placed in a cover is important to the style of the dining room. Not only does it add beauty in a third dimension, but it establishes a subtle finished quality to the table settings. Customarily, the simple folds are informal, tasteful, and quick to make. Sometimes a pocket for silverware is created in the napkin using a simple fold. Occasionally, a more formal feeling is desired and can be created with a complex fold. For instance, you might use a pleated fan fold, tucked into the water glass at a banquet, or a water lily fold in the center of the cover for a Victorian theme. There are dozens of different folds and variations and many uses for folded napkins (Figure 3-4).

Begin with large laundered, starched, and ironed napkins free of stains. Fold enough of the napkins to set the number of place settings in the dining room. Also fold extra napkins to reset the tables during the business hours and have them available at the sidestand. Four simple napkin folds are illustrated in Figures 3-5, 3-6, 3-7, and 3-9. An example of a table with dove-folded napkins is shown in Figure 3-8.

PREPARING THE SIDESTAND

A **sidestand** is a storage and service unit, sometimes with a computer terminal, located close to serving areas (Figure 3-10). A well-stocked sidestand eliminates the need for servers to make frequent trips to the kitchen for supplies or to place orders. One of the main opening duties is to stock the sidestand nearest your station

FIGURE 3-4 Use of a Decoratively Folded Napkin: A decoratively folded napkin can be used for many purposes. Photo by S. Dahmer

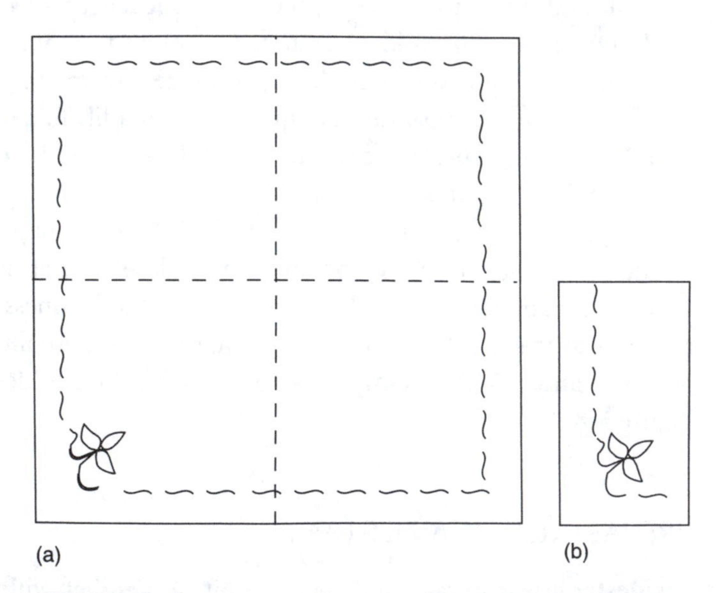

FIGURE 3-5 Oblong Fold: (a) Fold the flat napkin in half, then in half again to form a smaller square. (b) Fold the small square in half again lengthwise to form an oblong. (c) Place the oblong napkin in the cover with points (or decorative corner) to the left and out.

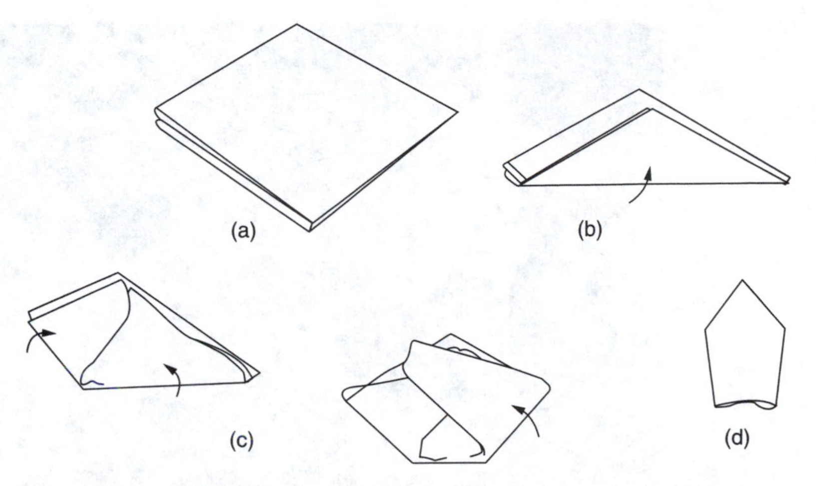

FIGURE 3-6 Crown Fold: (a) Fold flat napkin in half twice to form a square. (b) With loose points of square toward you, fold lower corner up not quite to top. (c) Fold both sides across center. (d) Turn over and place in center of cover with point toward center of table.

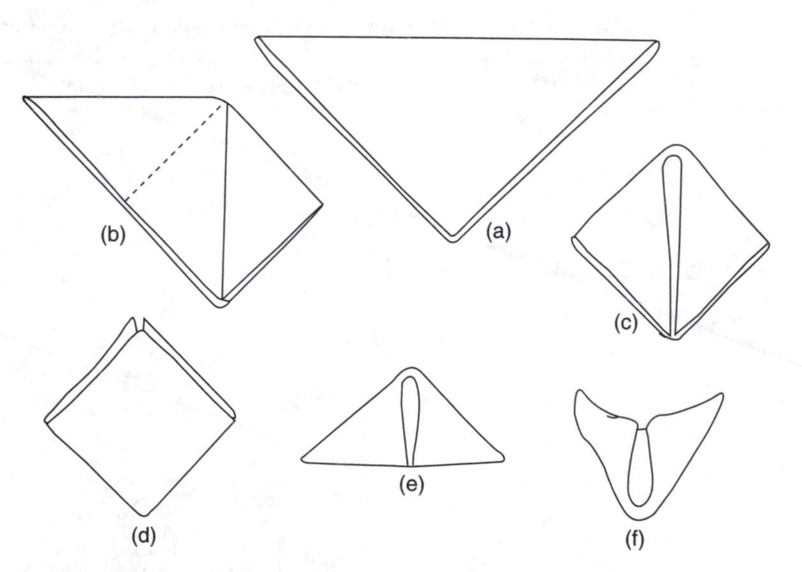

FIGURE 3-7 Dove Fold: (a) Fold the napkin in half to form a triangle with the point facing down, toward you. (b) Holding your finger at the center point of the top edge, fold the right corner down to the bottom point. (c) Repeat with the left corner. (d) Turn the napkin over so loose points are up. (e) Bring the bottom point up to meet the top to make a triangle. (f) Lift the napkin up at the center of the bottom, and stand the napkin up. Place the napkin in the cover with "wings" facing away from the guest.

FIGURE 3-8 Cover with Dove Folded Napkin: The dove fold is a simple napkin fold for beginners. The end result will hold its shape more easily if you use a heavy cotton napkin. Courtesy of PhotoDisc, Inc.

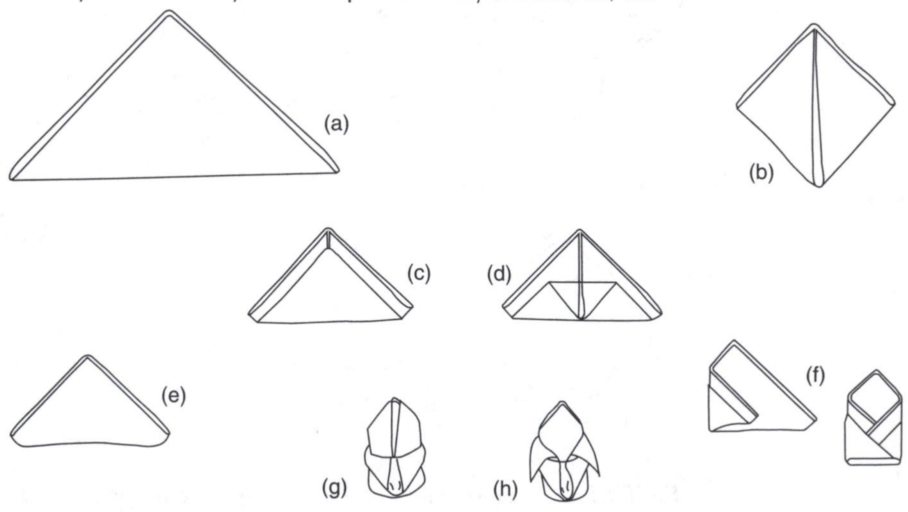

FIGURE 3-9 Bishop Hat Fold: (a) Fold the flat napkin in half diagonally to form a triangle with the point facing up. (b) Holding your finger at the center of the bottom, fold the right corner up to the top point. Repeat with the left corner, forming a square. (c) Bring the bottom point up to within one inch of the top point. (d) Fold the same point down to the center of the bottom edge. (e) Holding all folds in place, flip the napkin over to the other side with the point facing away from you. (f) Fold both sides to center, tucking right side into left. (g) Holding the tucked-in fold together, round out base so it stands up. Turn it around. (h) Gently pull down the front corners. Place in the center of the cover with the decorative side toward the guest.

FIGURE 3-10 Sidestand: A sidestand like this placed close to serving areas puts items needed in easy distance of the guests and reduces the number of trips that must be made to the kitchen.
Courtesy of Lakeside Manufacturing, Inc.

with serviceware, garnishes, beverages, and supplies. The items kept in stock at the sidestand vary among restaurants. Typical sidestand supplies include:

- Coffee warmers with fresh coffee
- Water pitchers, jugs, or carafes
- Clean, folded napkins
- Sponges and towels
- Order pads, guest checks, and extra pencils and pens
- **Condiments**, additives used to give flavor and relish to food, such as salt, pepper, steak sauce, ketchup, mustard, and horseradish in clean, filled containers
- Decorative garnishes and foods that complement the meal, such as lemon wedges, coffee cream, and jelly or preserves, according to the menu of the day, or tortilla chips or fortune cookies depending on the international theme of the restaurant (see Meal Accompaniments later in this chapter)
- Clean placemats
- Children's placemats, menus, bibs, and favors
- Silver and supplies for special food items, such as lemon squeezers, straws, iced tea spoons, and seafood forks

- Clean dinner menus and specialty menus, such as dessert and wine menus
- Drink coasters or napkins
- Tip wallets or trays
- China, silver, and glasses to set up covers

Because the sidestand is in plain view of the guests, it must be clean, neat, and presentable at all times. During the course of service, bussers should be notified to replace low inventories of supplies if you are too busy to replace them on your return trips from the kitchen.

Studying the Menu

WHAT IS A MENU?

The word ***menu*** means the food offerings of a restaurant and the actual printed or readable list of those foods. If readable, they are the individual lists on hard paper stock handed out to each guest. They may be the food list on boards on an easel, on the wall above the counter, or printed on the placemat. There are many styles of menus with food lists appropriate for a variety of different types of restaurants.

WHY STUDY THE MENU?

Familiarizing yourself with the menu for your restaurant should be one of your daily responsibilities. Study the menu so that you are aware of any changes in food choices and prices from day to day and to learn all of the menu items offered so you can sell foods that will be appealing to the guests and suggest side dishes. Know the menu so that you can answer the guests' questions. Following are detailed reasons why you should study the menu:

1. *Study the menu frequently because it may change occasionally or even daily*. Check with the chef if you see any new specials or menu items. Selections may change to give variety to the menu, and prices may change according to ingredient cost or seasonal availability of food items.
2. *Consider yourself a salesperson and the menu items your product*. Learn about the food on the menu by tasting all items at one time or another. Sell the food by describing the foods accurately and helping guests make satisfying choices. Knowing the menu helps you suggest side orders and build the check.

For instance, suggest items that complement the main part of the meal or additional foods that round out the meal or add to the festivity of the occasion, such as appetizers, champagne, or dessert.

3. *Be knowledgeable about food items so you can help guests who request information that may not be on the menu.* For instance, you may have to answer questions about vegetarian selections, budget-wise choices, kosher foods, food ingredients that may cause allergies, healthful choices, or foods for finicky eaters. With a knowledge of food ingredients and prices, you can help guests who have questions about the following:
   - *Vegetarian foods.* Know whether foods contain white meats or are without meat altogether.
   - *Budget-wise selections.* Know which items give guests the best value for their money.
   - *Kosher food.* Know whether any of the menu items are prepared under strict kosher rules and with kosher ingredients (see Content of the Menu later in this chapter).
   - *Ingredients that may cause allergic reactions.* Some guests may have an adverse physical reaction to foods because they have a sensitivity, or **food allergy**, to specific ingredients. The Food Allergy and Anaphylaxis Network states that a food allergy is an immune system response to a food that the body mistakenly believes is harmful. Guests who eat the foods they are allergic to may suffer a tingling sensation in the mouth, swelling of the tongue and throat, difficulty breathing, hives, vomiting, abdominal cramps, diarrhea, a drop in blood pressure, loss of consciousness, or even death. Symptoms of food allergies appear from within a few minutes to two hours after eating the offending food. Any food can produce an allergy, but according to the U.S. Food and Drug Administration, milk, eggs, peanuts, tree nuts, fish, shellfish, soy, and wheat produce about 90 percent of the allergic reactions. Guests with food allergies will usually ask you about those ingredients in menu items. To answer their questions, you should know the exact ingredients in menu foods and how they are prepared. If you are not absolutely sure, ask the chef.
   - *Healthful choices.* A guest may be diabetic and must know the sugar content in prepared foods. Some people may be on low-calorie, low-carbohydrate, low-fat, or salt-free diets and want to know how foods are cooked, whether they can get foods modified to their needs, or ask for sauces on the side so they may add them sparingly.
   - *Foods they may dislike.* Small children, for example, may not like foods such as onions, green peppers, or nuts.

Knowing the menu items in great detail will help you make suggestions and answer questions. If you do not know the ingredients in foods, do not guess. Ask the chef for the exact ingredients so that your guests can make informed choices.

TYPES OF MENUS

The most common menus are individual printed menus for breakfast, lunch, and dinner. Each meal usually has a separate menu, but occasionally a single menu contains selections for all three meals. The menu that is exclusively for luncheons contains sandwiches and light foods. The dinner menu (Figure 3-11) contains meals of larger portions, heavier foods like steaks and chops, and additional accompaniments such as vegetables. More food naturally causes the prices on dinner menus to be higher.

In addition to regular menus, there are menus for wines, called **wine lists** (see Wine Lists and Wine Charts in Chapter 8), and menus that sit up on the table, called table tent menus, which note particular items like specials, unusual drinks, or a dessert of the day. **Blackboard menus**, written on a blackboard at the entrance to some restaurants, are also in use. Servers in these restaurants memorize the menu and repeat it to guests from memory instead of issuing paper menus. There are also **banquet menus**, **early bird menus** (menus for guests who are willing to dine before a certain time), and menus without prices. There are **children's menus** with children's favorite foods, speedy service, child-size portions, and lower prices. These menus are very important for parents wishing to dine out with their children.

When food on a menu is **à la carte**, literally "from the card," it is listed as a single item and priced separately from other foods. An appetizer or salad listing is characteristic of this type of menu selection. When food on a menu is **table d'hôte**, meaning "table of the host," it is listed as a full-course meal and is priced as a unit, which may include rolls or other bread, soup or salad, meat/fish/seafood, potato or other starch, and vegetable for one price. The dessert is usually à la carte. Most menus contain an assortment of foods in both à la carte and table d'hôte listings.

CONTENT OF THE MENU

The content of the menu is organized into categories, or food groups, and also features specials. It is customary to list foods on a menu in the order in which they are usually eaten. Also, most items listed carry some designator indicating how they are prepared (see Methods of Food Preparation later in this chapter). The categories, specials, and trends in foods presented are discussed in the following sections.

Starters

SHRIMP & CRAB BISQUE
With corn and red peppers .... cup 3.95 .. bowl 4.95

BAKED FRENCH ONION SOUP
Swiss cheese crusted; garlic crouton 5.95

GRILLED CHICKEN AND PORTOBELLO MUSHROOM SKEWERS
Sesame ginger sauce 8.95

SPINACH AND ARTICHOKE FONDUE
Garlic toast points 8.95

FOUR CHEESE CRUSTED PORTOBELLO MUSHROOMS
Stuffed with fresh garlic spinach **8.95**

WOOD OVEN BAKED GOAT CHEESE
Tomato basil sauce, focaccia 8.95

WILD MUSHROOM PIZZA
Portobello and cremini mushrooms with swiss and asiago cheese 8.95

CRISPY FRIED CALAMARI
Cocktail and ranch dipping sauces **9.95**

WOOD OVEN ROASTED CRAB CAKES
Mustard mayonnaise 10.95

CRAB CRUSTED STUFFED SHRIMP
Roasted with garlic breadcrumbs 10.95

JUMBO SHRIMP COCKTAIL
Chilled gulf shrimp with zesty cocktail sauce and lemon wedges 10.95

ROASTED SEA SCALLOP SKEWERS
Bacon wrapped, apricot sauce 11.95

Salads

	SMALL serves 2 - 3	LARGE serves 4 - 5
HOUSE SALAD BOWL Mixed greens, hearts of palm, carrots, eggs, tomatoes, cucumbers, croutons; creamy ranch, 1000 island, house vinaigrette, blue cheese or French dressing	9.95	14.95
GOAT CHEESE AND FIELD GREEN SALAD Wood roasted vegetables, balsamic vinaigrette	9.95	14.95
CHOPPED TOMATO AND RED ONION SALAD Blue cheese dressing, crumbled blue cheese, and chopped lettuce	9.95	14.95
CAESAR SALAD Crisp romaine lettuce, parmesan cheese, and garlic croutons Add Char Crusted chicken.....3.00 planked salmon.....4.00 sliced tenderloin.....5.00	9.95	14.95
SPINACH SALAD Baby spinach, bacon, mushrooms, radishes, chopped eggs, warm mustard dressing	9.95	14.95
WILDFIRE CHOPPED SALAD Mixed greens, spit roasted chicken, avocado, tomatoes, blue cheese, bacon, scallions, corn, tortilla strips; tossed with citrus vinaigrette dressing	**11.95**	**16.95**
TERIYAKI CHICKEN SALAD Mixed greens, chopped vegetables, crispy tortillas, peanuts, cilantro, Asian vinaigrette	12.95	
GRILLED STEAK AND BLUE CHEESE SALAD Marinated sliced tenderloin, mixed greens, tomatoes, crispy onions, balsamic vinaigrette	14.95	

Sandwiches

Served with homemade cole slaw and french fries.

BLACK ANGUS BURGER
Char-grilled with lettuce and tomato 8.95
With Wisconsin cheddar, jalapeno jack, swiss, american or blue cheese 9.95

STUFFED BURGER OF THE DAY
Preparation varies; ask your server 10.95

DILL CRUSTED WALLEYE SANDWICH
Shredded lettuce, tomato, tartar sauce 10.95

PEPPERCORN TENDERLOIN STEAK SANDWICH
Oven dried tomatoes, lettuce, grilled red onions, and ranch dressing 13.95

PORTOBELLO MUSHROOM AND GOAT CHEESE
Field greens, roasted peppers and onions, balsamic vinaigrette 9.95

GRILLED CHICKEN CLUB
Grilled chicken breast, bacon, avocado, jalapeno jack, mustard mayonnaise, lettuce, tomato 9.95

TURKEY REUBEN
1000 island dressing, sauerkraut, swiss cheese 9.95

PRIME RIB FRENCH DIP
A classic sandwich served with melted cheese and au jus 12.95

D/MN 5/07

FIGURE 3-11 Sample Menu: A typical dinner menu contains many categories of related à la carte (individually priced) or table d'hôte (complete dinner) selections.

Courtesy of Lettuce Entertain You Enterprises, Inc.

Filet Mignon

Wildfire filets are cut from the center of a hand selected tenderloin and slowly aged for maximum tenderness and flavor.

FILET MIGNON
Broiled to perfection 28.95

BLACK PEPPERCORN FILET
Seared with spicy black peppercorns . . 29.95

BLUE CHEESE CRUSTED FILET
Broiled with our blue cheese crust 30.95

HORSERADISH CRUSTED FILET
OUR SPECIALTY! Wrapped in bacon and topped with a horseradish crust . **30.95**

BASIL HAYDEN'S® BOURBON TENDERLOIN TIPS
Marinated & grilled red onions 22.95

FILET MEDALLIONS OSCAR
Lump crab, asparagus, béarnaise sauce 28.95

WILDFIRE BONE-IN FILET
A very special cut 36.95

SURF & TURF
Filet and crab crusted lobster tail . . . . . . A.Q.

FOR THE LIGHTER APPETITE ~ TRY OUR PETITE FILET

Steaks and Chops

Wildfire steaks are aged for 21 - 28 days until they reach their ultimate tenderness and taste, brushed with our seasoning blend, and broiled to your preferred temperature.
Enhance your steak with one of our signature crusts, béarnaise sauce, or wood roasted mushrooms 2.00 each.

NEW YORK STRIP
Broiled to its fullest flavor 29.95

CHAR CRUSTED® RIB EYE
"Prime Rib" chop, served au jus 30.95

LAMB PORTERHOUSE CHOPS
Colorado's finest lamb 29.95

MUSHROOM CRUSTED PORK CHOPS
All natural farm raised **20.95**

ROUMANIAN SKIRT STEAK
Served with grilled red onions 22.95

PORTERHOUSE STEAK AU JUS
Two favorite cuts - New York Strip and Filet Mignon - in a single steak 30.95

BONE-IN NEW YORK STRIP STEAK
Served au jus. 32.95

STEAK ORDERING GUIDE

BLUE - Cold, Red Center • RARE - Very Red, Cool Center • MEDIUM RARE - Red, Warm Center
MEDIUM - Pink, Hot Center • MEDIUM WELL - Dull Pink Center • WELL - Broiled Throughout
We recommend that medium well and well done filet mignon be butterflied.
We are not responsible for steaks ordered well done.

Roasted Prime Rib of Beef

The "King of Roasts" is rubbed with fresh garlic, and slowly roasted.

PRIME RIB REGULAR CUT — 10 OZ WITH HORSERADISH SAUCE, AU JUS 22.95
PRIME RIB LARGE CUT — 16 OZ WITH HORSERADISH SAUCE, AU JUS 26.95
PRIME RIB "CHICAGO CUT" — 22 OZ WITH HORSERADISH SAUCE, AU JUS 30.95

Fresh Fish and Seafood

All roasted in our wood burning oven.

CEDAR PLANKED SALMON
Brown sugar soy glaze **20.95**

COCONUT SHRIMP
Ginger coconut sauce 20.95

SWORDFISH "LONDON BROIL"
Oven roasted tomatoes and red onions 22.95

EAST COAST SEA SCALLOPS
Fresh garlic spinach, lemon butter sauce 22.95

FRESH FISH OF THE DAY
Preparation varies, ask your server . . . . A.Q.

Above dinners served with a choice of redskin mashed potatoes, wild rice, french fries, or broccoli with herb butter.

· Barbecue rubbed sweet potato or giant baked potato add 1.95 ·
· Wildfire creamed spinach add 1.95 ·
· White cheddar au gratin potatoes add 1.95 ·
· Cheddar double stuffed potato add 2.95 ·

D/MN 2/07

FIGURE 3-11 (Continued)

Roasted Chicken

Served with redskin mashed potatoes, french fries, broccoli with herb butter or wild rice.

ROASTED HERB CHICKEN*
Marinated with a special blend of herbs and spices 15.95

BARBECUED CHICKEN*
Glazed and broiled with our special recipe barbecue sauce 15.95

LEMON PEPPER CHICKEN BREAST
Roasted in our wood burning oven **16.95**

*Sorry, all white meat not available

Barbecue and Combos

Served with homemade cole slaw, apple sauce and french fries.

BABY BACK RIBS
Smoked and grilled with our special recipe barbecue sauce .... half slab 15.95 . full slab 21.95

CHICKEN & BABY BACK RIBS COMBO*
Half a barbecued chicken and a half slab of our baby back ribs 21.95

HONEY MUSTARD GLAZED ST. LOUIS RIBS
Smoked and char broiled with honey mustard barbecue sauce ..... half slab **15.95** full slab **21.95**

CHICKEN & ST. LOUIS RIBS COMBO*
Half a barbecued chicken and a half slab of honey mustard glazed St. Louis ribs 21.95

RIBS & RIBS
A half slab of baby back ribs and a half slab of honey mustard glazed St. Louis ribs 21.95

*Sorry, all white meat not available

· Barbecue rubbed sweet potato or giant baked potato add 1.95 ·
· Wildfire creamed spinach add 1.95 ·
· White cheddar au gratin potatoes add 1.95 ·
· Cheddar double stuffed potato add 2.95 ·

Nightly Specials

Served with redskin mashed potatoes except where noted.
Quantities are limited.

MONDAY
Garlic Chicken
Oven roasted vegetables 15.95

TUESDAY
New York Strip Roast
Peppercorn sauce 19.95

WEDNESDAY
Wild Mushroom Stuffed Salmon
Red wine butter 20.95

THURSDAY
Half Long Island Duck
Sweet cherry sauce 19.95

FRIDAY
Dill Crusted Walleye
Wild rice, lemon butter sauce ....

SATURDAY
Filet Mignon Wellington
A classic; wild mushroom stuffi
shallot balsamic sauce ..

SUNDAY
Roasted Leg of Lamb
White cheddar au gratin potatoes, fresh garlic spinach

Potatoes and Ve

REDSKIN MASHED POTATOES ....
WILDFIRE WILD RICE ....
GIANT BAKED POTATO ....
BBQ RUBBED SWEET POTATO ....
WHITE CHEDDAR AU GRATIN POTATOE
STEAMED BROCCOLI WITH HERB BU
WILDFIRE CREAMED SPINACH ..
WOOD ROASTED MUSHROOM
CHEDDAR DOUBLE STUFFED PC

Wi

wildfirresta

FIGURE 3-11 (Continued)

Sp
A sp
server.

Categories

The foods on menus are grouped into categories according to the customs and preferences of management. Some of the more popular categories are listed here:

- An appetizer category includes foods served as a first course to stimulate the appetite of the guests. In a traditional restaurant, appetizers include small portions of specialty foods such as chicken skewers and stuffed mushrooms, and seafood items such as shrimp and clams. The appetizer category in less formal restaurants is often dominated by finger foods and foods of ethnic origin or influence, such as nachos, meant to be shared by several guests at the table.
- Soups may be placed in a separate category, grouped with appetizers or salads, or included with table d'hôte entrées. Soups may be clear, cream-based, hot, or cold.
- Salads, usually lightly tossed fresh greens, are grouped with soups, by themselves in a category, or featured with the main part of the meal.
- **Entrées**, the main part of the meal in American service, vary extensively and can be grouped in any number of individual categories. Some of the more common categories are steaks, seafood, meats, poultry, pastas, sandwiches, entrée salads, and specialties. Generally, a vegetable and some type of carbohydrate accompany the meat/fish/seafood entrées, and sometimes an entrée is featured by itself.
- The dessert category usually includes pies, cakes, cheesecakes, ice cream, and specialties such as crème brûlée and tiramisu. The dessert usually completes the meal.
- The category of beverages includes coffee, tea, milk, lemonade, juices, soft drinks, and other drinks. Cocktails and wines may also be listed on the food menu.

The types of foods and the number of selections in each category vary from one restaurant to another. Exclusive restaurants list gourmet-type foods, and family restaurants list homestyle foods. Restaurants may list calories of each item, feature kosher foods, or identify low-fat or low-salt foods for healthful selections. Some restaurants still feature a large number of selections; others have followed the current trend toward limiting choices to cut costs (see Sample Menu in Figure 3-11).

cials

cial of the day may be attached to the menu or described to guests by the A special may be a chef's specialty, a regional dish, or a seasonal food in

ample supply. A chef's special is a dish the chef prepares exceptionally well. Walleye in Minnesota and gumbo in Louisiana are regional specials. Fresh strawberries or melon and some seafood are seasonal specials. The price of a special may be lower if it is a seasonal item. As a server, you should strive to describe specials in an appetizing way to increase sales.

Trends

Current trends affect the content of the menu. The National Restaurant Association reports that Asian appetizers, organic or locally grown produce, specialty sandwiches, whole-grain breads, pan-seared items, exotic mushrooms, fresh herbs, pomegranates, and free-range or grass-fed meat choices are popular. There is also a trend toward downsizing desserts into mini desserts that can be eaten in a couple of bites. Specialty coffees like espresso-based lattes, cappuccinos, mochas, and Americanos are also very popular, and bottled water is often chosen over tap water.

A restaurant may be set up to serve kosher foods and will indicate that it is a kosher restaurant on the menu. **Kosher foods** are permitted to be eaten by people of the Jewish faith who observe kosher dietary laws. Kosher does not refer to a particular cuisine, but to a set of rules regarding preparation of foods with kosher ingredients in a kosher facility (or kitchen). Food that is kosher must be supervised during preparation and made with ingredients that are approved by certifying supervisors. The kosher consumer extends beyond the Jewish community to include Seventh-Day Adventists, Muslims, vegetarians, and many health-conscious Americans. People with allergies to pork or shellfish select kosher foods to avoid allergic reactions to those ingredients. Foods certified kosher and **pareve** are foods containing neither meat nor dairy products, but under Jewish law may contain a very small amount of milk so may not be appropriate for guests who are allergic to milk.

The increasing ethnic diversity of the United States will accelerate the trend toward more ethnic restaurants and ethnic dishes on the menus. Guests will enjoy unique menu choices such as Szechuan Chinese, Hunan Chinese, German, Greek, Japanese, Tex-Mex, Cajun-Creole, Mexican, Thai, Mandarin, Vietnamese, Indian, Mediterranean, Latin, Asian, and Italian.

METHODS OF FOOD PREPARATION

Because guests often ask how foods on the menu are prepared, you should know the more common preparation methods, as follows:

- **Baked.** Cooked by dry, continuous heat in an oven
- **Boiled.** Cooked in liquid at the boiling temperature of 212°F at sea level, so that bubbles rise to the surface and break

- **Braised.** Browned in a small amount of fat and then cooked slowly in a little liquid in a covered pan
- **Broiled.** Cooked by direct heat, either under the source of heat, as in a broiler, or over the source of heat, as on a grill
- **Fried.** Cooked in hot fat. Pan-fried and sautéed means cooked in a small amount of fat. Deep-fried means cooked while immersed in a large amount of fat.
- **Grilled.** Cooked over direct heat, usually hot coals
- **Pan-broiled.** Cooked in a heavy frying pan over direct heat, using little or no fat
- **Poached.** Simmered in enough liquid to cover the food
- **Roasted.** Cooked uncovered without water added, usually in an oven
- **Sautéed.** Browned or cooked in a small amount of hot fat
- **Simmered.** Cooked gently in a liquid over low heat just below the boiling point
- **Steamed.** Cooked in steam with or without pressure
- **Stewed.** Simmered slowly in enough liquid to cover the food

PREPARATION TIME

Preparation time is the time required to cook and dish up a food item on the menu. The length of time depends on the equipment in the kitchen, the efficiency of the chef, and the number of orders already placed by other servers. Preparation times can best be learned by experience. Once you know them, however, you will be able to time your orders competently.

Some of the more common food preparation times are as follows:

- Eggs: 3–5 minutes
- Fish, fried or broiled: 10–15 minutes
- Liver: 10–15 minutes
- Chateaubriand: 30 minutes
- Steak, one-inch thick
  - rare: 10 minutes
  - medium: 15 minutes
  - well-done: 20 minutes
- Lamb chops: 20 minutes
- Pork chops: 15–20 minutes
- Game: 30–40 minutes
- Chicken: 10–20 minutes
- Soufflé: 35 minutes
- Pasta: 10 minutes

New equipment and preparation methods have shortened preparation times considerably. Some foods are precooked in advance and heated to serving temperature when ordered. Other foods are prepared early in the day and kept at serving temperature constantly, either on a steam table, if served hot, or in a refrigerator, if served cold. Equipment such as the microwave oven shortens the preparation time of food items to minutes and seconds. For the guests' convenience during rush hours, know which items can be served immediately.

MEAL ACCOMPANIMENTS

Meal accompaniments are condiments, decorative garnishes, and foods that complement the entrée (Figure 3-12). Part of your responsibility may be bringing the

FIGURE 3-12 Garnished Food: Meal accompaniments include decorative garnishes. It may be the responsibility of the server to add the garnishes to the guest's meal.
Courtesy of PhotoDisc/Getty Images

condiments to the table to complete the order and adding the garnishes and complementary foods to an entrée prepared by the chef. Make sure the garnishes are attractive and that condiment containers are clean. Some accompaniments may be kept at the sidestand for convenience. Examples of meal accompaniments are as follows:

- Lemon wedge with fish
- Tartar sauce with fish
- Ketchup with hamburger
- Mustard with hot dogs
- Steak sauce with steak
- Applesauce with potato pancakes
- Hot syrup with pancakes
- Dressing with salads
- Butter and jam with bread and rolls
- Crackers with soup
- Salsa with tortilla chips
- Clarified butter with lobster
- Parsley and other greens to add color to an entrée
- Cream, sugar, and artificial sweetener with coffee
- Lemon, sugar, and artificial sweetener with tea
- Lemon or lime in ice water

Closing the Dining Room

Servers have closing duties to perform between luncheons and dinners and at the end of the day. When closing after lunch, reset enough tables for the anticipated number of dinner guests.

At the end of the day, take time to close the dining room properly. Leave your station, sidestand, and kitchen area in a clean, orderly condition, ready for business the next day. The following are some closing duties:

- Make sure the mouths of ketchup bottles are wiped clean and cruets are grease-free.
- Remove all salt, pepper, and sugar containers and place them on trays. Wipe the containers with a clean, damp cloth and refill them. Containers should be periodically emptied and washed.
- Take all condiments, such as ketchup, mustard, and steak sauce, to the refrigerator.

- Foods in packages, such as individually wrapped crackers, nondairy creamers, and jellies, maintained in sound condition and remaining in baskets or containers at guests' tables, can be used again, depending on the policies of management. Discard the remaining packaged butter, cream, and wrapped foods that have been contaminated at guests' tables.
- Strip the tables of their tablecloths. Tables should not be set up for the next day, because settings gather dust. Instead, assemble all of the serviceware on trays for use the next morning, and cover them with napkins.
- Empty all coffee containers and have coffee equipment cleaned.
- Return unused and voided numbered guest checks to the supervisor.
- Reconcile receipts and cash with computer totals if using a computer system (see Closing at the End of the Day in Chapter 7).
- Turn off all heating equipment, such as roll warmers and coffee-making equipment.

KEY TERMS

Sidework
Mise en place
Station
Receptionist/Reservationist
Taking reservations
Silencer
Cover
Table tent menus
Sidestand
Condiments
Menu
Food allergy
Wine lists
Blackboard menus
Banquet menus
Early bird menus
Children's menus
À la carte
Table d'hôte
Entrées
Special
Kosher foods
Pareve
Baked
Boiled
Braised
Broiled
Fried
Grilled
Pan-broiled
Poached
Roasted
Sautéed
Simmered
Steamed
Stewed
Preparation time
Meal accompaniments

REVIEW

1. Define station, sidestand, sidework, silencer, and special.
2. List several methods of assigning stations to servers and the reasons why different methods of assignment are used.
3. How can a reservationist minimize the chance for no-shows?
4. Recommend a procedure for keeping the sidestand replenished on a routine basis.
5. Discuss the advantages and disadvantages of cabinet doors for concealing the shelves of sidestands.
6. Why is studying the menu necessary?
7. List special problems people may have with certain foods on a menu.
8. List various ethnic menus or foods on a menu that are of ethnic origin.
9. Compare the job of a salesperson in a retail store with the job of a server. In what ways are their jobs alike? In what ways do they differ?
10. What are the main differences between the luncheon and the dinner menus?
11. Discuss several reasons why restaurants have specials on their menus. Why should servers suggest specials to their guests?
12. Why should a server be familiar with methods of food preparation and preparation times?
13. Why may preparation times vary from one restaurant to another?

PROJECTS

1. List the opening duties for your restaurant or a local restaurant, and divide them among the servers who work the morning hours. You may want to rotate groups of duties so the waitstaff can be responsible for a variety of jobs on different days.
2. Ask a trainee to demonstrate how to place a tablecloth on a table in a professional manner. Repeat this procedure with the center items on the table. Then have the trainee demonstrate how to remove the tablecloth with and without center items on the table. Have all trainees try this procedure.
3. With clean, starched, and pressed napkins, practice the four folds shown in this chapter until you can complete them without looking at the directions. Research other, more complex napkin folds and folds to hold silverware, and demonstrate them to others.
4. Study several menus from local restaurants. Identify the following parts:
   a. Those items that are à la carte
   b. Those items that are table d'hôte
   c. The various groupings of foods and the number and kind of selections in each group
   d. The special attachment or the place where a special may be attached
5. Study a menu and consult with a chef in order to do the following:
   a. Identify the main ingredients of each food item.
   b. Describe the method of food preparation for each food item.
   c. Identify the preparation time necessary to cook or prepare each food item.
   d. Name the meal accompaniments that go with each food item.
6. Have a chef discuss how restaurant foods may be adapted to guests' special diet needs as they request.
7. List all of the closing duties you can think of or that must be done in your restaurant if you are already employed. Post these in the kitchen, and make a regular habit of checking them off at the end of the day.

CASE PROBLEM

A Reservation Error

A guest reserved a table for nine people at 7:00 P.M. The reservation server became very sick and had to leave. She did not record the reservation before she left for the day. The restaurant was very busy, and a server who was assigned to handle reservations had to greet the guests who did not have a table reserved for nine people at 7:00 P.M. Answer the following questions:

- What could the server say to help reduce the disappointment of the guests?
- How can this situation be resolved so the guest who thought his reservation had been recorded is satisfied with the way the situation is handled?
- What steps can be taken to prevent this sort of error from happening again?

Chapter Four

Initiating the Service

Objectives

After reading this chapter, you will be able to:

- ✔ Understand how to seat guests to control the traffic flow in the dining room.
- ✔ Discuss how to approach guests who are seated.
- ✔ Identify techniques for taking orders.
- ✔ List many questions to ask guests who are ordering.
- ✔ Describe two techniques for taking orders.
- ✔ Identify four methods of taking orders.
- ✔ Discern between appropriate and inappropriate topics of conversation.
- ✔ Prepare to answer common questions guests may ask.
- ✔ Know how to make suggestions and increase the size of the guest check.
- ✔ Manage the timing of the service and meal.
- ✔ Identify the methods of placing orders in the kitchen.
- ✔ Describe ways servers are notified that their orders are ready to be served.

Initiating the service means beginning the service. It includes greeting and seating the guests, taking the orders, timing the meal, and placing and picking up the orders in the kitchen. Appropriate conversation, answering questions, and making suggestions to increase sales are also important parts of the initial service.

Seating Guests

WHO SEATS GUESTS?

Guests may be greeted and seated by the maître d'hôtel, the host, or the head server, who keeps track of open tables, assigns waits, and seats people as their turn or reservation comes up. This process can make guests feel immediately welcome and give them a good first impression of the restaurant. It also allows the maître d'hôtel or host to control the traffic flow of guests in the dining room by seating guests evenly among stations and staggering the seating. Some restaurants use seating management software on their computer to help the host track reservations and seating (see Restaurant Reservations and Table Management in Chapter 7). Guests are allowed to select their own tables, booths, or counter spaces in some restaurants.

WHERE TO SEAT GUESTS

Common sense dictates where parties of guests should be placed in a dining room. Utilize tables according to party size. For example, seat a large family at a large round table and a couple at a smaller table for two, called a **deuce** (Figure 4-1).

Public health is protected by law. Clean indoor air statutes prohibit smoking in public establishments in many states. Other states limit smoking to designated smoking areas of restaurants. These areas must be set aside from nonsmoking areas and posted for smoking. Seat smokers in smoking sections and nonsmokers away from them in nonsmoking sections.

Loud, noisy parties may be placed in private rooms or toward the back of the dining room so they do not disturb other guests. Elderly guests or guests with disabilities may wish to be near the entrance to minimize walking distances. Young couples like quiet corners and good views. Of course, if guests request a specific location, you should try to accommodate them.

FIGURE 4-1 Table for Two: Seat two people at a table for two called a deuce. Photo by S. Dahmer

HOW TO SEAT GUESTS

Approach guests with words of greeting, such as "Good evening." Guests will inform you when they have reservations. When they do not have reservations, ask them, "How many are in your party?" and, if appropriate, "Would you like to be seated in a smoking or nonsmoking area?" When there appears to be one person, ask, "Table for one?" instead of "Are you alone?" If guests must wait for seating, take a name or provide a coaster pager (see Guest Paging in Chapter 7), and tell them you will notify them when a table becomes available. If there is dining seating available, take clean menus and lead the guests to the table.

When women are in the party, a maître d'hôtel or head server seats one or more of them in the seats with the best view (Figure 4-2). The men in the party usually assist in seating the other women present. A female host or server generally pulls out the women's chairs to indicate where they may be seated, but does not actually seat them. When patrons will be placed at wall tables with banquette seats on one side, the tables may be pulled away from the seats by the maître d'hôtel or host so that guests may be seated easily. An open menu is presented to each guest, and friendly conversation is carried on throughout this initial service. At this time,

FIGURE 4-2 Woman Being Seated: The maître d'hôte or headserver seats the women in a party of guests, although the men in the party may assist some of the women themselves.
Courtesy of Hennepin Technical College, Eden Prairie, Minnesota; Photo by S. Dahmer

remove table settings that will not be used, bring booster chairs or high chairs for children, and supply missing serviceware. Fill water glasses or have them filled by a busser.

CONTROLLING SEATING

To control the traffic flow in the restaurant, the greeter should avoid seating two groups of guests in the same station at the same time. Instead, parties should be seated in different stations so that any one server is not overburdened and the guests receive better service. The greeter should also stagger the seating of large parties so that servers will have equal opportunities to serve large groups without having too many of these groups at one time.

On a busy day, many restaurants are so popular that guests have to wait for a table. The waiting areas can be mini-destination areas, where guests enjoy first courses and beverages before moving into the dining room for the main course. The greeter should be sure guests are seated in order of their arrival and registration. Guests with reservations should be given seating preference at their appointed times.

Approaching the Guests

The server should approach the guests after they have had time to look at the menu. Approach the host of the party first, because he or she may wish to order for the guests. The host is the spokesperson and will address the server for the entire group. He or she is the person most attentive to the welfare of all the members of the party and often takes the seat at the head of the table. Approach the host first from the left, and if he or she does not wish to order for the others, begin taking the order from the next person to the right.

When approaching guests to begin service, greet the party of guests with a friendly, appropriate greeting such as "Hello" or "Good evening." Some restaurant managers like you to announce your name and that you will be the server for the meal. Inform the guests of unlisted specials and your recommendations at this time, and then ask, "Would you like to order a cocktail or beverage?"

If you are busy with another table when the newest party of guests is seated in your station, approach the new group and tell them you will be with them shortly. They will appreciate your attention and be patient knowing you will be with them as soon as possible.

Taking Orders

TECHNIQUE OF TAKING ORDERS

Stand erect to the left of the guest with the order pad supported in the palm of your hand and a sharp pencil ready (Figure 4-3). Never place your book or order pad on the table to take an order. Use one of the two following techniques to help you identify the first person who orders and to know where to start serving when you bring the food:

1. *Make a mental note of the first person who orders.* If you are taking the order on paper rather than on the guest check, you may write down a unique characteristic identifying the first person. For example, note hair color, glasses, clothing, or tie. Then from that reference person, proceed taking orders counterclockwise around the table. When you serve the meal, you can serve exactly what each guest ordered without asking questions.
2. *Note and circle the seat number of the first person who orders.* Seat numbers should be understood in advance. For example, the seat on the kitchen side of the table can be known as seat number one, the seat to the left of this seat going counterclockwise is seat number two, and so forth. If all servers use this system,

FIGURE 4-3 Taking the Order: When taking the order, the server stands erect and holds the order pad in the palm of the hand. Using a sharp pencil makes orders easy to read.
Courtesy of PhotoDisc, Inc.

any other server or restaurant employee can deliver food to your table when you are busy.

Write clearly and systematically for your own benefit and that of the kitchen staff, who must prepare the order exactly according to your instructions. Take the order completely. A few of the many questions you may have to ask the guests are as follows:

- Whether drinks should be iced
- Choice of salad dressing
- Choice of vegetable or side
- How meat should be cooked
- Sour cream or butter on baked potatoes
- How eggs should be cooked
- When to serve coffee

Use common abbreviations known to kitchen staff when you take orders. Some common abbreviations are shown in Figure 4-4. To prevent error, you may repeat

Tuscan Chicken Soup - Ch soup	Filet Mignon - F M
Mixed Greens Salad - Mx greens	Rare cooked - r
Caesar Salad - Czr	Medium rare cooked - mr
Hamburger - Hamb	Medium cooked - m
French Fries - ff	Medium well cooked - mw
Chicken Scallopini - Ch scal	Well cooked - w
Cedar Planked Salmon - C P Salm	Tiramisu - tira
Strip Steak - Strp stk	Key Lime Crostata K L pie

FIGURE 4-4 Example of Menu Abbreviations: Servers and kitchen staff should agree on menu abbreviations. Using abbreviations speeds up writing and interpreting the order.

the order back to the guests for their confirmation, especially when the order is given in an irregular fashion. Retrieve the menu from each guest after you write his or her order.

METHODS OF TAKING ORDERS

There are four methods of taking orders in restaurants:

1. A **checklist order system**. With a checklist order system, the server simply selects the food choice from a preprinted list of menu items (Figure 4-5). The server indicates the quantity and size of each item and multiplies quantity times unit price to get the total for that item in the right-hand column. The right-hand totals are added for the guest check total at the bottom. This system is used in short-order, fast-food establishments with a limited menu.
2. A **guest-check order system**. With a guest-check order system, the server hand-writes the order directly on the guest-check form (Figure 4-6). The bar order is written on the back, and the food order is recorded on the front of the check, or the bar order may be taken on a separate check. The food check is placed in the kitchen and filled by the chef in turn with other orders. It is returned to you when you pick up the meal from the kitchen. The bar total is added to the food total, and the check is presented to the guest for payment. Taking the order directly on the guest check is most satisfactory when the menu is printed with numbered meal combinations such as breakfasts and lunches. This method is

SERVER	STATION		CHECK NO.
			119953

Qty	Item		Amount
1	BIG STEER	PLATTER P (L) S C	7.79
	HAMBURGER	PLATTER P L S	
	CHEESEBURGER	PLATTER P L S	
1	FISH SAND.	PLATTER P (L) S	5.99
	TEXAN	PLATTER O	
	DANDY BOY	PLATTER L S T	
	BREADED TENDERLOIN	PLATTER L S T	
	GRILLED CHEESE	PLATTER	
	CHARBURGER	PLATTER O	
	BLT SANDWICH	PLATTER L M T	
	BLT CLUB	PLATTER L M T	
	TUNA SALAD	PLATTER L M	
	GRILLED CHICKEN	PLATTER L M T	
	FISH TAIL SANDWICH	PLATTER L S	
	GRILLED TENDERLOIN	PLATTER	
	CHICKEN	SEAFOOD COMBINATION	
	SPAGHETTI	HAM STEAK	
	FISH DINNER	PORK CHOPS	
	BROILED FISH DINNER	CHICKEN STRIPS	
	SHRIMP DINNER	BROILED CHICKEN	
	CHICKEN FRIED STEAK		
	C.B. GROUND BEEF STEAK	10 oz STRIP SIRLOIN	
	COUNTRY FRIED STEAK	12 oz T-BONE	
	6 oz RIBEYE	PRIME RIB	
	8 oz RIBEYE	PORK SPARE RIBS	
	OE O OM OW H SCR	HAM BAC SAU STK	
1	choc sundae		2.99
1	cherry pie		3.29
	CHEESE | H & C | WEST. | SPAN | SUPREME		
	HOT CAKE 1-2-3 | B & G		
	SALAD BAR		
	TOSSED SALAD | BC | TI | FR | RA | OV		
	SLAW | COTTAGE CHEESE | APPLESAUCE		
	CHEF SALAD | LG | SM | SUPREME | LG		
	FF | ONION RING | B.P.		
2	(COLA) | DIET | (SPRITE) | ROOT B. | COFFEE		2.98
	MILK LEMONADE TEA C-MILK HOT CHOC		

THANK YOU - PLEASE COME AGAIN

FIGURE 4-5 Checklist Order Form: A server needs to indicate the quantity or size of each item selected and extend the amount of the charge on a checklist order like this.

best used when the kitchen is a single unit, so that the check does not have to be excessively handled.

3. A **notepad order system**. With a notepad order system, the server makes a chart on a blank pad of paper. Each food course is listed at the heading of a column, and guests are noted down the left margin. The choices that the guests make are then noted under the appropriate headings (Figure 4-7). A glance at a column quickly indicates the entire party's choices of cocktails, salads, entrées, or other

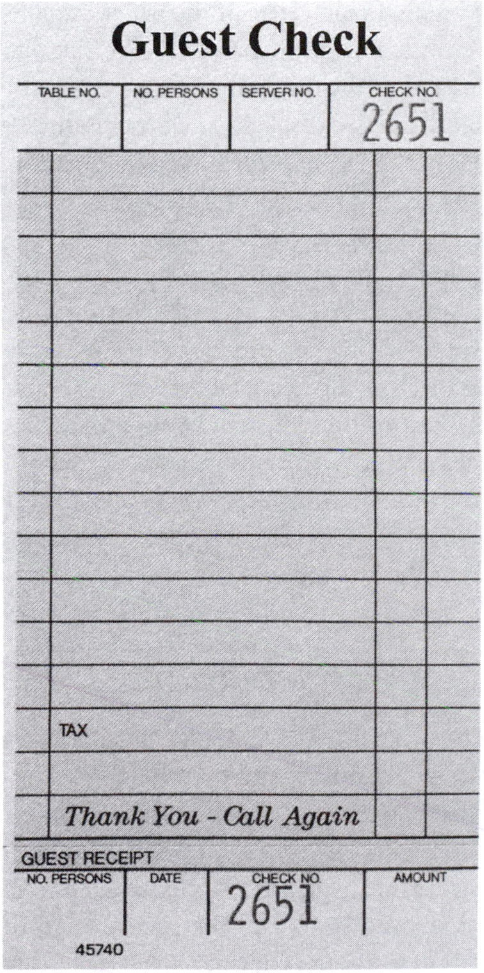

Guest Check

TABLE NO.	NO. PERSONS	SERVER NO.	CHECK NO. 2651

TAX

Thank You - Call Again

GUEST RECEIPT

NO. PERSONS	DATE	CHECK NO. 2651	AMOUNT

45740

FIGURE 4-6 Guest Check: The server writes the order neatly and legibly on a guest check form like the one shown here. Courtesy of PhotoDisc/Getty Images

		Cock	Appet	Salad	Entree	Veg	Pot
Red hair	1.	House chard.	—	Spin Warm mustard	Salmon	Broc	Wild Rice
	2.	Apple Mart. up	—	House Honey mustard	Lamb chop m	Mixed veg.	Bak
	3.	Manh. rocks	Calamari	House roq.	FM-mr	Broc	Bak
	4.	Sam Adams	—	Field greens Bal. Vineg.	Coconut Shrimp	Broc	Redskin

FIGURE 4-7 Example of an Order on a Pad of Paper: A server-written order may be taken on a pad of paper in chart form, as shown here.

food courses. Carbon copies are made, parts of the order are rewritten, or the order is keyed into the computer for the kitchen staff (See Taking Orders Using a POS System in Chapter 7), and the original order is retained by the server, who uses it to serve cocktails, appetizers, salads, and other items. After the meal, the server summarizes the order on a guest check or prints a guest check on the computer and presents the check to the guest for payment.

A notepad system of taking the order is advantageous when the guests order full-course meals and the servers are responsible for plating and/or garnishing some courses, such as soups, salads, and desserts, and serving them in the proper order. This method of taking the order is also used with a multiunit kitchen consisting of separate chefs for steam table foods (stews, soups), grilled foods (eggs, steaks, chops), and cold foods (salads, appetizers, desserts).

4. **Handheld computer order system.** With a handheld computer order system, an order is taken directly on a handheld computer by touching prompts or writing on the screen with a stylus (see Handheld Order Terminals in Chapter 7). The order is sent electronically to the kitchen and bar.

Appropriate Topics of Conversation

Brief conversation with guests, or small talk, is a pleasant part of your job and makes the guests comfortable in restaurant surroundings. Keep comments and topics positive in nature; you can always find something nice to say. Small talk may include complimentary comments about menu items, food suggestions, and opinions about the weather. Avoid any negative comments, such as those that may concern coworkers or the restaurant owners and the specifics of problems in the kitchen. Also resist the temptation to discuss your personal life or that of the guest. Keep small talk short and pleasant while attending to your work as a server. If guests are busy talking among themselves, attend to your business of serving without entering into the conversation.

Answering Questions

As a server, you are asked questions about food, the restaurant, the community, and even the state. Prepare yourself to answer questions by reading and listening. Be knowledgeable enough to answer questions intelligently and "suggestive sell" (discussed in the next section). You should know the following information about your restaurant:

- The hours, phone number, and address of the restaurant
- The restaurant's Web site

- The menu and the ingredients in every dish served. Know the special of the day. Also know the general menu of other dining rooms and coffee shops on the premises and the hours they are open.
- Historic facts of interest about your restaurant and community
- Events and attractions in the community, area, and state

If you are asked a reasonable question that you cannot answer, get the correct answer from someone who knows. For example, when a guest asks whether there are onions in the beef bourguignonne and you do not know, say, "I don't know, but I'll ask the chef."

Many restaurants have a staff meeting at about 11:00 A.M. or 4:00 P.M. so that the host or maître d'hôtel can go over the menu for the day and announce changes in service. At this meeting, you are informed of the specials, soup selection, and dessert assortment, as well as about large groups that have reservations. In other restaurants, food and service changes are posted for you to read when you come to work.

Making Suggestions and Increasing Sales

Suggestive selling is an extremely important task of servers in a restaurant. **Suggestive selling** means recommending the restaurant's food, beverages, and services to guests to improve their restaurant experience and to increase the size of the guest check. Larger guest checks translate into increased profits for the restaurant and larger tips for you. Happy guests become regular patrons.

Suggest cocktails before the meal and perhaps cold or hot appetizers to enjoy with cocktails. Ask, "Ice water or bottled water?" Suggest side orders that complement the entrée, such as ham with omelets and mushrooms with steak. Also suggest beverages, desserts, and after-dinner cordials. Suggesting from dessert tables is particularly advisable, because the mouthwatering display of desserts should almost sell themselves. Some dessert displays are on trays or mobile carts that can be wheeled directly to the guests' tables for presentation. If you know that guests are celebrating, suggest the wine list or a birthday dessert or cake if your restaurant offers one. Another service is to suggest bringing an extra plate to divide a regular portion between children or to suggest children's portions or menus.

During a special training session or periodic sessions in conjunction with menu changes, the manager or trainer may actually let you taste the foods served in the restaurant. This is a wonderful way to become acquainted with foods so that you can answer any questions and make recommendations. When asked to help a guest choose between two menu selections, say why you would recommend one. Do not degrade the second choice. Another way of helping a guest select is to describe the preparation and accompaniments of each, thus letting the guest decide according to the additional information.

Avoid being overbearing or pushy about suggesting foods and beverages. Suggest only in a helpful way, and be sure your suggestions are appropriate to the meal. Be specific when you suggest a food. Questions such as "Will there be anything else?" and "Would you like dessert?" are too general. Ask specifically whether the guests would like menu items such as the crème brûlée or the strawberry cheesecake. When they ask you what is good today, reply with a specific suggestion instead of saying, "Everything is good." Or you might say, "Everything is good here, but my personal recommendation for today is _____."

As you master the art of making specific suggestions, you can whet the guests' appetites by using appropriate adjectives that tempt the palate. For instance, say, "Would you like to begin your meal with our crispy tostada appetizers topped with roast pulled pork, cilantro, lime, and onions?" or "May I suggest our wonderful signature dessert, coconut cream pie with brûléed bananas and caramel?" Or you may have an opportunity to describe a food item in appetizing terms such as "Our roast pork is made of the finest choice pork tenderloin with a maple fig demi glace" (Figure 4-8).

FIGURE 4-8 Dessert: To increase sales, whet the appetite of the guest by describing specific foods, such as this dessert, in mouthwatering terms. Courtesy of PhotoDisc, Inc.

Timing the Meal

The server has the responsibility to time the entire meal so that the pace of the meal is smooth, comfortable, and neither rushed nor delayed. You are the sole communication link between the guests and the kitchen. If guests indicate they are in a hurry, guide them to menu items that can be prepared quickly instead of rushing the chef.

After taking the complete order, the server must decide when to place it in the kitchen. A good rule of thumb is to submit the entrée order just prior to serving the appetizer. Hold the order for a short time when you see that the guests are lingering over cocktails.

In a single-unit kitchen, the chef sees that the entire order is ready at the same time. The only responsibility of the server is to submit the order as soon as possible. In a multiunit kitchen, the server coordinates the meal and submits the order in accordance with the length of time necessary to prepare the entrées. The meal order may have some entrées that take more preparation time than others (see Preparation Time in Chapter 3). Submit these orders in separate stages so that they are ready at approximately the same time. For example, knowing that pork chops and a medium steak take 15 minutes, a chef's salad takes 10 minutes, and beef burgundy is ready immediately, submit the grill order first, the salad order 5 minutes later, and the steam table order last. By placing the orders in this fashion, they are ready simultaneously, ensuring that hot foods are hot and cold foods are cold. In larger restaurants today, an expeditor coordinates all of the orders coming out of the kitchen. Dessert orders should be submitted and picked up immediately after the meal.

Placing Orders in the Kitchen

The method of communicating orders to the kitchen staff varies among restaurants, depending on the size, type of kitchen, type of service, and availability of a computer system. There are three methods of communicating orders to the kitchen:

1. *Spoken.* In some restaurants, you orally communicate the order to the kitchen by entering the kitchen and clearly giving the order to the proper chef, who may write down the order.
2. *Written.* In some restaurants, the order may be written on a checklist or guest check and given to the chef, who can arrange all of the orders to be filled in sequence. Occasionally, in restaurants with larger or multiunit kitchens, servers

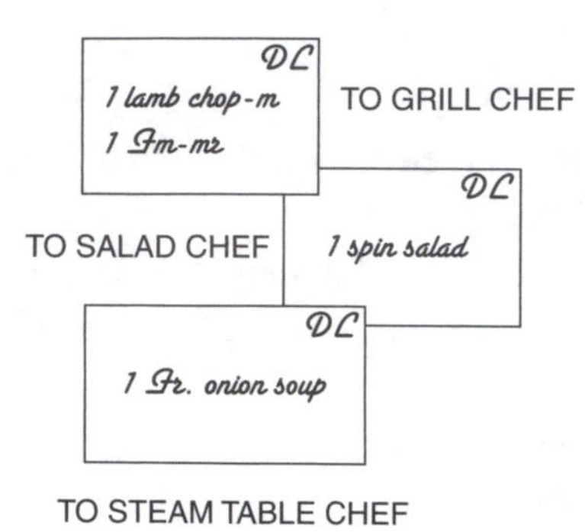

FIGURE 4-9 Example of Separate Orders to Kitchen Prep Areas: In some restaurants, servers have to rewrite parts of an order for the various food preparation areas in the kitchen.

use notepads and must rewrite parts of the order for the separate kitchen areas, as described previously. Separate kitchen area orders are shown in Figure 4-9.

3. *Entered.* In other restaurants, the order is keyed into a computer terminal and sent to the chef electronically (see Getting Orders to and from the Bar and Kitchen in Chapter 7).

Some forgetful servers put orders back in their jacket or apron pockets and neglect to submit them. Regardless of the method you use, chefs do not prepare orders until they receive them.

FIGURE 4-10 Pager: One of the several ways in which a server is notified that an order is ready is by pager. Courtesy of PhotoDisc, Inc.

Picking up Orders from the Kitchen

There are various ways in which you may be notified that your order is ready to be picked up from the kitchen. A lighted number on the wall of the dining room may indicate that your order is ready, or the chef may take the responsibility of orally notifying you. Some servers wear **pagers**, electronic devices that signal them by beeping or vibrating, when their orders are ready (Figure 4-10).

Compare the food with the order to see that the order is complete. Be sure the food is attractively arranged and garnished. Remedy any of your mistakes or those of the chef at this time. Arrange the plates on the tray so that they are well balanced to carry to the table.

KEY TERMS

Deuce
Checklist order system
Guest-check order system
Notepad order system
Handheld computer order system
Suggestive selling
Pagers

REVIEW

1. Which restaurant personnel are responsible for seating guests?
2. What determines the smoking policy in a restaurant?
3. Why is it advantageous to seat guests instead of allowing them to seat themselves?
4. What are some of the clues that may help you identify the host of a party of guests?
5. Describe one or more methods to help you remember the items each guest ordered.
6. Why is the notepad order system most popular for dining rooms with table d'hôte selections on the menus?
7. Why do order methods differ from one restaurant to another?
8. What topics are appropriate for conversation with guests?
9. What suggestions would you make in the following instances?
   a. A guest orders apple pie.
   b. The guests have ordered cocktails before their meal.
   c. A couple is celebrating their anniversary.
   d. A guest orders a club sandwich.
10. Suggest the following menu items in appetite-whetting terms:
    a. Bruschetta
    b. Monte Cristo sandwich
    c. Veal and Pork Bolognese
    d. Cedar-planked salmon
    e. Baked Alaska
11. Explain the procedure for timing the following entrée meal order in a restaurant with a multiunit kitchen: chateaubriand, red snapper, lamb chops, and lasagna.

PROJECTS

1. Role-play seating the guests. Include conversation with the guests, placement of parties in the dining room, removal of extra serviceware, provision of water, and other services to make the guests comfortable.
2. Meet with service and kitchen personnel and agree on abbreviations for food items on the menu. Make a list and post it in the kitchen. Or, as a trainee group, make a list of common abbreviations from a menu. Discuss the necessity of uniform use of abbreviations by all personnel.
3. Practice taking orders from fellow trainees posing as guests. Be sure to include suggestive selling.
4. Describe the order-taking method used in your restaurant. Or, as a trainee group, select a familiar restaurant and discuss its order-taking methods.
5. Investigate community events and attractions that would interest your guests. Get information from city hall, the chamber of commerce, service clubs, newspaper, radio publicity, and historical societies.
6. Go to restaurants of various sizes and types and ask service personnel how they place and pick up orders in the kitchen. Bring the information to class and discuss it with other trainees.

CASE PROBLEM

Initiating the Service

A female executive is seated with her male client in your station at 1:00 P.M. She tells you they are celebrating the culmination of a large business deal, and they are ready to relax after a month of hard work. They are discussing the menu, and the executive is considering ordering a shrimp cocktail (an appetizer) as her lunch. Her guest is considering ordering a pasta entrée and a side salad selection. A topic of conversation is calories and how they both like to maintain healthful eating habits. They seem to have plenty of time to enjoy their lunch. Answer the following questions:

- What signs do you look for to identify the host? In this case, whom would you approach?
- What words would you use to initiate service? How would you use suggestive selling?
- How would you time this meal?
- To whom would you present the check?

Chapter Five

Serving the Meal

Objectives

After reading this chapter, you will be able to:

- ✔ Learn the proper technique for serving each course at tables and booths.
- ✔ Understand how to serve beverages.
- ✔ Describe how to clear soiled dishes.
- ✔ Identify techniques for serving efficiently.
- ✔ Identify guests with special needs and how to best meet those needs.
- ✔ Handle complaints that may arise.
- ✔ Describe how to handle the guest check and payment at the end of the meal.
- ✔ Define how to increase, receive, and report tips.

Serving means bringing the food and beverage order to the table and attending to the guests' needs throughout the meal. Good service involves serving foods and beverages in an efficient manner that combines proper serving techniques and courteous attention to the guests. It also means being prepared to handle unusual circumstances during the course of service. Serving includes the suitable handling of the payment and tip.

Serving Tables and Booths

SERVING EACH COURSE

Women and elderly people are served first, out of traditional courtesy, unless the party consists of a host entertaining friends. In that case, begin with the guest of honor at the host's right. Continue serving around the table counterclockwise. If you have taken the order correctly, you will never have to ask a guest what he or she has ordered as you serve.

Serve all foods, such as appetizers, soups, salads, entrées, and desserts, from the left of the guest with your left hand (Figure 5-1). Support the dish with your fingers beneath the plate and your thumb on the rim, and place it in the center of the cover. Using your left hand may take practice if you are right-handed, but the left-handed serve eliminates the possibility of elbowing the guest.

Each course is served as follows:

1. The **appetizer** is served first and is offered to arouse the appetite and set the tone of the meal. It is a small portion of hot or cold food, such as lump crab cake, oysters on the half shell, or pâté with crackers. In a traditional restaurant, this first course is placed on a small plate called an **underliner** and centered before the guest (Figure 5-2). A seafood fork is brought with shrimp and oysters. It is either placed on the right-hand edge of the underliner or inverted into the bowl of the outside spoon of the table setting. In less formal restaurants, appetizers like chicken skewers, bruschetta, or nachos may be served on small plates or in baskets.
2. Soup may be served in place of the appetizer or as a second course. It is served in a small, handleless cup on a saucer. A soup spoon must be provided if it is not furnished with the table setting. Serve the soup in the center of the cover, with the soup spoon on the right side of the saucer.
3. The salad is the next food served. It is placed to the left of the cover, allowing space in the center for the entrée. Salad dressing may be served in several ways. It may be served on the individual guest's salads at the table by the server, brought to the table in a self-serve container, added to the salads in the kitchen according to the order, or served in small, individual containers on the side as requested by the guests. Fresh black pepper in a pepper mill or parmesan cheese and a cheese grater may be brought to the table and, if the guest agrees, added to the salad from the guest's left.
4. The entrée is the main part of the meal in American service. Prior to serving the entrée, check to be sure the table is properly set for the entrée selected. For

FIGURE 5-1 Serving Guests from the Left with Left Hand: All foods should be served from the left of the guest, and dishes should be placed with the server's left hand.
Courtesy of Hennepin Technical College, Eden Prairie, Minnesota; Photo by S. Dahmer

example, with steak, add a steak knife, and with lobster, add lobster crackers and a seafood fork. Dinner rolls may be served with the salad course or, at this time, as an accompaniment to the entrée. Condiments such as steak sauce should be brought to the table when requested. Serve the entrée in the center of the cover. When serving a meat entrée, place any bone or fatty part away from the guest and the meaty portion toward the guest to ensure that the first bite will be pleasing. Refill the water glasses.

FIGURE 5-2 Underliner: An underliner with a doily can be used for butter, cream and sugar, salad dressing, and sauces that are passed around at the table.
Photo by S. Dahmer

5. The dessert is the last course served. A dessert fork or spoon should be brought with the food item and placed to the right of the dessert plate. Replenish coffee and water at this time. The service is complete when the guests ask for or you bring their check.

SERVING BEVERAGES

Give maximum attention to beverage service throughout the meal. Ask the guests if they would prefer ice water or bottled water and serve the water after guests have been seated. Take the drink order next, before the appetizer order. Table wines are typically served with the main part of the meal but are opened and served when ordered (see Taking the Wine Order in Chapter 8), as are all other beverages. Coffee and tea are served after the entrée, with dessert, after dessert, or as requested by the guests. Keep glasses refilled with ice water, coffee cups filled with coffee, hot water in teapots, and wine glasses filled with wine until the bottle is empty. Suggest an after-dinner drink when you take the dessert order.

Serve all beverages, such as water, milk, coffee, tea, and alcoholic beverages, from the right of the guest with your right hand (Figure 5-3). Place beverages such as milk or tea to the right of the cover, and refill beverages such as coffee or wine without lifting the cup or glass from the table. When pouring a hot beverage for

FIGURE 5-3 Pouring Beverages from the Right with the Right Hand: When pouring beverages, the server should leave the glass or cup on the table and pour with the right hand from the right of the guest whenever possible. Photo by S. Dahmer

guests seated close together, use a clean, folded napkin in your left hand to shield the guests from the hot container and alert the guest that you are about to pour.

Coffee Service

Although there are many specialty coffees, regular coffee and **decaffeinated coffee**—coffee with only a trace of caffeine—are most commonly served with a meal. Coffee is often one of the last foods guests linger over, and it leaves a lasting impression as they finish their restaurant experience. Good coffee is made with scrupulously clean coffeemakers, a proper brewing technique, and good-quality coffee and water. Coffee should be served at approximately 160°F (71.1°C). The cup and saucer should be brought and placed to the right of the cover with the handle to the right. Bring or place the teaspoon to the right or on the rim of the saucer. Cream, sugar, and sweetener should be available as accompaniments. A good way of remembering whether a guest has ordered regular or decaffeinated coffee is to place the spoon on the rim perpendicular to the table edge if regular coffee was ordered, and slightly angled inward if decaffeinated was ordered. Some servers place a paper coaster on the saucer under the cup for decaffeinated. Pour the coffee slowly without touching the rim of the cup. Leave room for milk. Some restaurants' policy is to serve full cups or mugs of coffee.

Tea Service

Tea is an aromatic beverage made by infusing tea leaves in boiling water. Teas range from unfermented green teas to lightly fermented oolong to fully fermented black teas. There are also herbals, Chai, and jasmine teas. Teas are served hot or cold. For hot tea, a **tea sachet**, an individually bagged portion of loose tea, is placed in a small individual ceramic or stainless pot with enough hot water at about 185°F (85°C) for a pint of tea. When tea is extracted to the strength desired, the guest will remove the sachet to prevent overextraction. Iced tea is often served with, not after, the entrée. Iced tea is served in tall, narrow glasses over ice. Sugar, sweetener, and lemon slices should be provided as accompaniments.

SERVING AT BOOTHS

To determine how to serve guests in a booth, assume you are standing at the end of the booth; serve the guest farthest from you first with the hand farthest from the guest's seat. This means that the guests on your right would be served their food with your left hand; the guests on your left would be served with your right hand. Serving in this manner prevents contact with the guest (Figure 5-4). However, always serve at the convenience of the guest, even if you must break the rules of proper service. For example, you may have to pick up the cup and pour coffee for a guest seated deep in a booth when serving properly would be awkward.

CLEARING DISHES FROM THE TABLE

Clear the dishes when all guests at the table have finished the course. You can often tell they are finished because the fork and knife are placed parallel to each or the table knife blade is inserted into the tines of the fork on their plate. When in doubt, ask whether they are finished. Then remove all soiled dishes and utensils completely before serving the next food item. Clear dishes from the right of the guests with the right hand. Again, move from guest to guest in a counterclockwise direction around the table. In addition to dishes, pick up all dirty glasses and soiled silver.

FIGURE 5-4 Serving Guests in Booths or Close Quarters: When serving guests seated at booths or in close quarters, serve each guest with the hand farthest from the guest. Serve guests at the server's right with the left hand, and, as pictured, serve guests at the left with the right hand.
Courtesy of PhotoDisc, Inc.

Clear soiled dishes to a nearby tray on a tray stand. Work quietly and efficiently, and never scrape or stack the dishes at the table. Stack the dishes on the tray so they are well balanced and safe for you to carry to the kitchen (see Safety: Preventing Accidents in Chapter 6).

Before dessert, you should remove all serviceware except water glasses and coffee cups, and crumb the table. **Crumbing the table** is the process of sweeping loose food particles into a clean plate to make the table more presentable. This can be done with a clean, folded napkin or with a **crumber**, a pencil-length metal tool with a groove in it for sweeping the table clean.

Efficiency When Serving

MINIMIZING STEPS

Time is important when serving guests, and you can save time by minimizing steps whenever possible. Never walk back to the kitchen empty-handed. Take dirty dishes back to the kitchen on your way to pick up food. Have a tray stand available when you come out with your next course. By cutting down the time spent in serving guests, you not only increase the number of people you can serve but also improve your efficiency. An increase in the number of guests served and faster service increases your tips.

MAINTAINING FOOD QUALITY

Serve hot foods hot and cold foods cold. Pick up and serve foods in the order that will maintain this temperature quality. For example, when serving several tables, pick up salads first and hot soups or cold ice cream last on the same tray. Likewise, serve the hot soup or cold ice cream first and the salad last.

ATTENDING TO GUESTS

There is no excuse for ignoring a guest. Allow little delay between courses, and keep your eye on the guests as you serve others in your station. Guests indicate they need you by a look, gesture, or remark, and you should respond promptly to their needs.

Special attention should be given after you have served the entrée. When the guests begin to eat the main course, check back to be sure that all entrées are as ordered, satisfactory, and complete. Mistakes can be remedied easily at this time, and the guest will not be angry or dissatisfied. The end of a meal is too late to adjust a complaint. Replenish rolls, water, and coffee quietly. Give guests an appropriate amount of attention. Beware of giving too much attention by hovering over guests, monopolizing the conversation, and constantly interrupting.

GIVING SPECIAL AMENITIES

Guests often want to take home food that they ordered but were unable to finish eating. The restaurant will provide plastic bags, Styrofoam boxes, white cartons, or some other type of container for this purpose. The server should remove the guest's plate to a sidestand or kitchen to box the uneaten portion for the guest. Write the name of the food item and date on top of the container for the guest. Another option is to give guests the containers so they can box remaining food themselves.

Another service is to present a mint to each guest at the table when you bring the check. Some restaurants may instead present a complimentary biscotti, fortune cookie, after-dinner wine, chocolate, or some other form of appreciation for the guest's patronage. Many restaurants give a free dessert to patrons who have a birthday or anniversary. Servers may even gather around the guest(s) of honor and sing an appropriate celebration song.

Handling Unusual Circumstances

YOUR BEHAVIOR TOWARD ALL GUESTS

A good server strives to serve all guests equally well. Thousands of satisfied guests are necessary to run a restaurant successfully, and a server cannot be particular about which guests he or she serves. Most guests appreciate your efforts, but some are difficult to please. Handle each situation, no matter how unusual or unpleasant, with genuine interest in serving the guest the best way you know how.

HANDLING GUESTS WITH SPECIAL NEEDS

Occasionally, servers encounter guests who have special needs, such as guests who are very young, disabled, foreign-born, or intoxicated.

Very Young Guests

If your assigned party of guests includes a young child, suggest a high chair or booster seat, if appropriate. Place the high chair at the table out of the aisle. Do not attempt to secure the child in the high chair or booster seat yourself. It is the parent's responsibility to be sure the child is secured so that the child will not fall out and sustain an injury. Treat the child as an important person. Be patient and pleasant, and assist the parents in making the child comfortable. Without being obvious, move the sugar, salt, pepper, and breakable items out of the child's reach.

Some restaurants have children's menus; however, never take a child's order without consulting the parents. Do not fill glasses too full. Use low dessert dishes and tumblers or covered plastic glasses instead of stemmed glassware. Parents appreciate extra napkins, bibs, novelty placemats, crayons, and other favors for the children (Figure 5-5). Bring small children some food, such as breadsticks or crackers, as soon as possible, because they are not patient. Cheerfully warm an

FIGURE 5-5 Dining with Children: When dining out with children, parents appreciate favors, crayons and paper, children's placemats, crackers, children's meals, and covered drink containers to make the experience more enjoyable. Photo by S. Dahmer

infant's bottle when asked, but return it warmed to the parent, not the child, so the parent can test it for a safe temperature. Children can cause accidents if they are allowed to run around the dining room. If children are playing in the aisles or disturbing other guests, suggest to the parents that they keep the children at the table to prevent injury to the children.

Disabled Guests

Occasionally, a person who is disabled comes into a restaurant alone. A **disabled person** is one who has a physical or mental impairment or challenge as a result of conditions that are congenital or acquired by heredity, accident, injury, advanced age, or illness. Be attentive to his or her needs. The guest will tell you how he or she would like to be helped. Understanding the disability and assisting him or her properly and discreetly helps the guest enjoy the meal. For example:

- A person in a wheelchair may wish to be pushed up to the table, but be sure the wheelchair is out of the aisle. This guest may need assistance with the salad bar.

- A person who is visually disabled needs a lot of attention, but be careful not to be offensively oversolicitous. Hang up the guest's coat and belongings and gently lead the guest to a seat. Discreetly move barriers to the area he or she requests. Quietly ask whether you may acquaint the guest with the menu. Then let the guest select the meal—the guest will select easily handled items. Assist the guest with the salad bar. Do not fill glasses or cups too full. As you serve, inform the guest where the food and beverage items are being placed and whether a plate itself is hot.
- A person who is hearing impaired may be able to give you an order verbally like any other guest. Some, however, prefer to write out or point to their choices. Be alert to the fact that people who are hearing impaired speak with hand movements. If you are concerned about an accident, gently touch the person on the right or left shoulder to indicate that you are serving from that side.
- A person who is mentally disabled may or may not be able to communicate their needs depending on their particular challenges. Some mentally disabled people will have family or staff with them to help them order and pay. Others will be able to handle the order and payment themselves. Be attentive and patient and assist them as needed.

Foreign-Born Guests

Guests who speak a different language may be unable to communicate their order to you without some difficulty. Try to determine some choices by similarity of words or hand gestures. Suggest, as best you can, that they point to the items on the menu that they desire.

Intoxicated Guests

Guests who are intoxicated should be seated in the back of the dining area or in a private dining room where they do not disturb other diners. Sometimes a guest who is intoxicated is bothersome or rude. Under no circumstances should you serve an intoxicated guest more alcohol in your establishment (see The Concern about Serving Alcohol in Restaurants in Chapter 8). Suggest nonalcoholic drinks, food, and coffee. Be tolerant; call the manager if the situation goes beyond your control. Be sure the guest remembers to pay the bill.

HANDLING COMPLAINTS

Complaints arise when guests do not get the food and service they think they deserve. In general, the better the table service and food quality, the fewer the complaints.

Do not take offense, however, when guests do complain. Valid complaints are the restaurant's feedback and should be used to improve service for those who come to the restaurant in the future. Allowing a complaint to be ignored or handled poorly will result in a disgruntled guest, who will then talk negatively about the restaurant.

Some complaints can be prevented before they occur. For example, if a guest orders a separate side order for which there is a charge, mention the extra charge for this specialty so that there is no surprise when the guest finds it on his or her check. When an order is delayed in the kitchen, reassure the guest that the order has not been forgotten. Remain professional and never place blame on other employees for a mistake or a delay, especially when the table is under your control.

By all means, avoid arguing with guests. Use tact and courtesy, and respect their opinions. Remember, the guest is always right. A good procedure for handling complaints is as follows:

1. Listen to the details of the complaint and take time to understand it.
2. Restate the complaint briefly to show you understand.
3. Agree truthfully to a minor point (e.g., "Thank you for bringing that up"). This puts you in the position of empathizing with the guest and looking at the complaint from the guest's point of view.
4. Handle the complaint promptly. Make an immediate adjustment or correction if you can. Treat your guests as you would like to be treated if you were in their position.

The Guest Check and Payment

When the guests indicate they have finished, complete the check and present it to the left of the host. This may be done in one of two ways:

1. *The check may be placed face down on the table.* Thank the guest and make brief departing remarks like "Please pay the cashier" and "Come back again soon." The guest then pays the cashier.
2. *The check may be presented in a folding wallet, and the guest pays you* (Figure 5-6). Take the wallet, guest check, and payment to the cashier or server computer terminal to complete the transaction.

If guests pay in cash, never ask if they would like change. Bring the change or offer to do so. It is the guest's privilege to say, "Please keep the change." Then thank the guests and invite them back again.

FIGURE 5-6 Presenting the Guest Check: When guests have finished their meal, bring the guest check. It may be presented in a folding wallet, as seen here, and the guest pays you. Courtesy of PhotoDisc/Getty Images

Payment is often made by personal check, traveler's check, or credit card instead of cash, although many restaurants will not accept personal checks today. If your establishment still accepts checks, accept a personal check for the amount of the guest check (plus any additional amount for a tip the guest may wish to add). Read all parts of the check to be sure it is accurate and complete (Figure 5-7). Ask

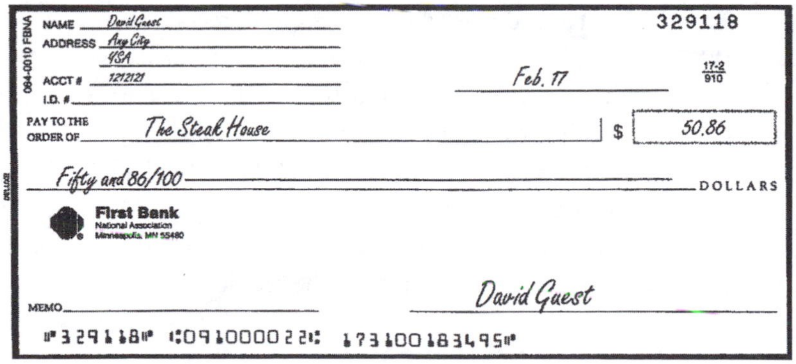

NAME David Guest
ADDRESS Any City
USA
ACCT # 1212121
I.D. #
329118
Feb. 17
17-2
910
PAY TO THE ORDER OF The Steak House $ 50.86
Fifty and 86/100 DOLLARS
First Bank
National Association
Minneapolis, MN 55480
MEMO
David Guest

FIGURE 5-7 Personal Check: Examine all parts of a personal check carefully to make sure the date, restaurant name, both numerical and written amounts, and the signature are correct. With practice, this can be done quickly.

FIGURE 5-8a Example of a Credit Card: Credit cards may be used, instead of cash, to pay for a restaurant meal. Courtesy of PhotoDisc/Getty Images

for acceptable identification, such as a driver's license, compare the picture and signature, and have the check authorized by your supervisor. Take precautions to avoid accepting a bad check.

Credit cards such as Visa, MasterCard, Diner's Club, American Express, and Discover are also used extensively in our mobile society. These cards should be signed by the cardholder. There are several ways to use the card. You may put the card through a card reader at the computer terminal and print the guest check with the computer printer (see Completing Each Transaction in Chapter 7), or place it in the addresser or stamping machine with the credit slip on top and slide the bar over both to imprint the slip. List the costs of the meals, tax, and bar total on the slip and subtotal the amount (Figures 5-8a, 5-8b, and 5-8c). Adding on a tip is the cardholder's decision. Bring a pen, and have the guest read, total, and sign the slip (Figure 5-8d). Then compare the signature with the one on the credit card to be sure they are identical, and return the credit card.

Receiving the Tip

A **tip** or **gratuity** is a monetary reward for courteous and efficient service. Guests are not obligated to leave you any tip, but tipping is traditional. Tips are incentives to do a good job (Figure 5-9). If service is very good, the tip will usually be good. But

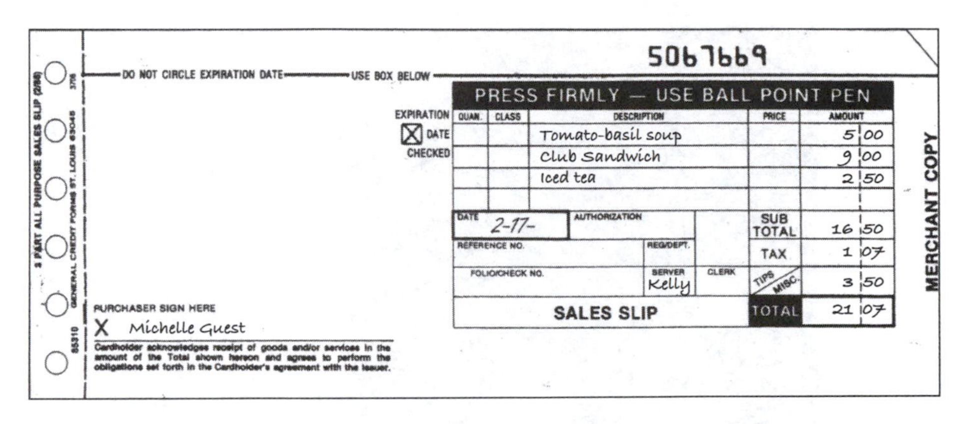

3 PART ALL PURPOSE SALES SLIP (2/88) GENERAL CREDIT FORMS ST. LOUIS 63045 3735 55310

DO NOT CIRCLE EXPIRATION DATE — USE BOX BELOW

5067669

PRESS FIRMLY — USE BALL POINT PEN

EXPIRATION ☒ DATE CHECKED

QUAN.	CLASS	DESCRIPTION	PRICE	AMOUNT
		Tomato-basil soup		5 00
		Club Sandwich		9 00
		Iced tea		2 50

DATE 2-17-

AUTHORIZATION

REFERENCE NO.

REG/DEPT.

FOLIO/CHECK NO.

SERVER Kelly

CLERK

SUB TOTAL	16 50
TAX	1 07
TIPS / MISC.	3 50
TOTAL	21 07

SALES SLIP

PURCHASER SIGN HERE

X Michelle Guest

Cardholder acknowledges receipt of goods and/or services in the amount of the Total shown hereon and agrees to perform the obligations set forth in the Cardholder's agreement with the issuer.

MERCHANT COPY

FIGURE 5-8b Example of a Credit Card Slip: The credit card is imprinted on a sales slip like the one shown here.

FIGURE 5-8c Credit Card Authorization: To authorize the credit card transaction, a server slides the bar over the credit card and sales slip to imprint the slip.
Courtesy of PhotoDisc, Inc.

FIGURE 5-8d Guest Signing Credit Card Slip: The credit card slip is presented to the guest at the table, and the guest signs the slip to authorize the payment.
Courtesy of PhotoDisc/Getty Images

FIGURE 5-9 Guest Leaving a Tip: The guest leaves a tip as a monetary reward for courteous and efficient service. The size of the tip often depends on how well the guest has been served.
Photo by S. Dahmer

sometimes tipping is based on the quality of food instead of the attention given by the server. Servers should remedy poor food quality with the kitchen so that it does not reduce the size of their tips. A good server should not worry about a regular guest who is pleased with service but does not tip well, or at all. Continue to give the steady guest your best service, because regular guests can give the restaurant a good recommendation and repeat business.

Tipping is very important, because a server's base pay is low, and tips make up the difference in earnings. Generally, the size of the tip is between 10 and 20 percent of the total amount of the guest check. Although the amount is at the guest's discretion, the following is a well-accepted guideline for tipping:

- 10 percent for poor service
- 15 percent for average service
- 18–20 percent for good service
- More than 20 percent for extraordinary service

A tip may be given to you in various ways. If it is handed to you, thank the guest politely. If it is left on the table, pick it up before the table is cleared. It if is added to the charge slip by the guest or added automatically to the check by the restaurant, you will receive it from the cashier or in your paycheck.

If several servers share the responsibility of one table, they should divide the tip. In some restaurants, servers pool their tips, then divide them equally at the end of the shift. It may also be policy to share a percentage of tips with the bussers, drink runners, expeditors, and other assistants.

INCREASING THE TIP

There are many ways to increase the tip or to be deserving of a generous tip. Better service, greater sales, and increased repeat business all add up to higher income from tips for you. Here are a few ways to increase your tips:

- Be neat in your appearance.
- Give guests friendly greetings.
- Be friendly, helpful, and efficient.
- Smile often when appropriate.
- Check often to see whether guests are in need of service, and offer to help them. You might say, "Is everything all right?" or "How is your food?" If there is a problem, take care of it immediately.
- Serve orders to guests as soon as possible.
- Offer appropriate condiments with foods, such as steak sauce with steak, tartar sauce with fried fish, lemon with baked fish, and ketchup with French fries.
- Make suggestions and merchandise the menu to every guest. Tell guests about specialties of the house. Use mouthwatering words to describe foods. Suggest appetizers, side orders, special beverages, and desserts the guest may enjoy. This will increase the check, and with that, the tip (see Making Suggestions and Increasing Sales in Chapter 4).
- Pour water and coffee for guests as needed.
- Thank guests as they leave and say "Come again" or another appropriate good-bye.

REPORTING TIPS

Under federal law, all tips count as taxable income. You must report all tips you receive to your restaurant manager and to the Internal Revenue Service for paying income, Social Security, and Medicare taxes.

Servers earning $20 or more in tips in any one month must report the amount to their employer in writing. This includes tips earned in cash, on credit cards, and those tips earned indirectly in tip pools. Fill out a report by the tenth day of the month following the month you earned the tips. Your report should include your name, address, Social Security number, and the total amount of tips you received during that period. Keeping a record of your daily tips will be a good defense in proving the total tips earned in case of an audit.

KEY TERMS

Serving
Appetizer
Underliner
Decaffeinated coffee
Tea
Tea sachet
Crumbing the table
Crumber
Disabled person
Tip or gratuity

REVIEW

1. Define crumbing and appetizer.
2. Briefly describe the differences between serving food and serving beverages.
3. What should you do if a guest asks you to take his or her plate before everyone at the table is finished with the course?
4. Why should you avoid scraping and stacking dishes at the table in front of the guest?
5. How should you handle the complaint when a guest says:
   a. "The food is cold!"
   b. "My steak is too rare!"
   c. "The chili is too spicy!"
   d. "You spilled coffee on me!"
   e. "My table is too close to the smoking section."
6. How can you protect a guest from the hot coffee pot when you are pouring coffee at the table?
7. Using the least number of coins and bills, indicating the amount of change to be returned for each of the following transactions:
   a. Given $5.00 for a $4.27 check.
   b. Given $10.00 for a $7.56 check.
   c. Given $15.00 for a $13.22 check.
   d. Given $20.03 for a $12.78 check.
8. What is the procedure for accepting personal checks to pay the tab?
9. What size tip is usually given for good service?
10. Why is it true that you can increase your tips with fast, efficient service?
11. How should you react when a guest does not leave you a tip?

PROJECTS

1. Role-play serving and clearing dishes from a table. Take turns being the guest and the server.
2. Observe servers when you are a guest in a restaurant, and note errors made by the service personnel. Discuss these points with other trainees.
3. Brainstorm ways to save time and effort in the serving process.
4. Discuss unusual circumstances concerning guests, other than those mentioned in this chapter.
5. Have one trainee learn and demonstrate the charge card transaction.
6. Interview a good server, and find out what he or she would suggest to increase tips.
7. Discuss how to handle the following situations:
   a. The electricity is off, and there are items that the guest cannot order.
   b. The wine cellar is locked, and the staff cannot find the key.
   c. A child drops his plate of food on the floor or an adult spills his drink (consider safety issues).
   d. You, the server, drop a tray of food.

CASE PROBLEM

Handling a Complaint

A party of several guests is celebrating together. One guest orders a food item from the menu. You, the server, go to place the order and find the food item is no longer available. You apologize and explain the situation to the guest, but the host of the party complains to you about the situation. Answer the following questions:

- In this situation, list the steps you would take to avoid a conflict with the host of the party.
- In this case, you were not at fault, but how could you have known about the problem before it occurred?
- How can you, as the server, avoid disappointing guests as they dine at your station in the future?

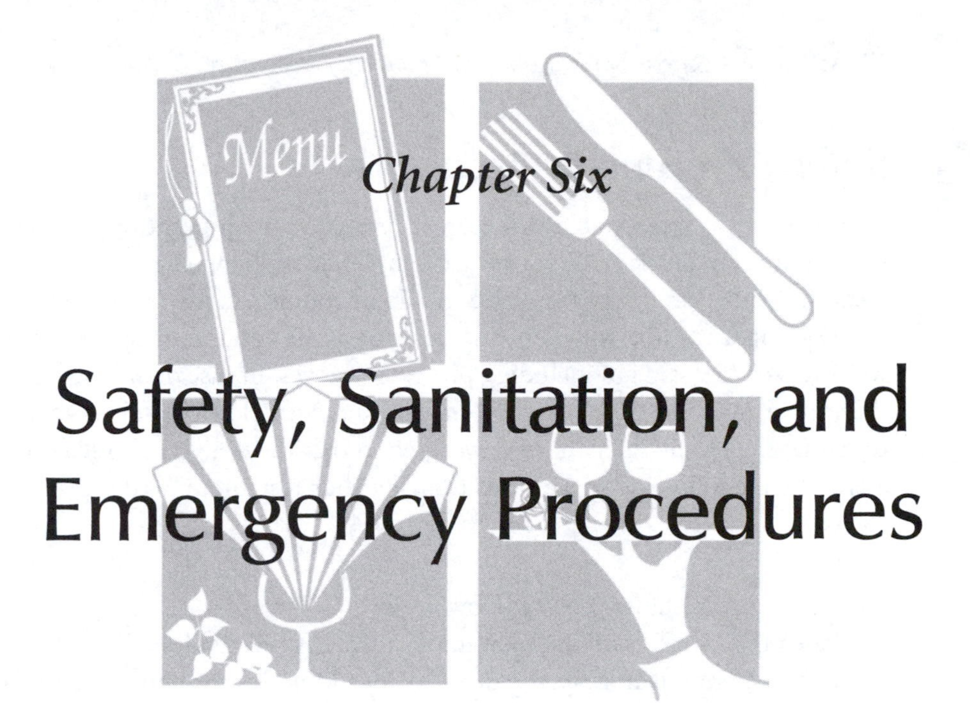

Chapter Six

Safety, Sanitation, and Emergency Procedures

Objectives

After reading this chapter, you will be able to:

- ✔ Recognize and remedy safety hazards in a restaurant.
- ✔ Identify and ensure safe food handling practices.
- ✔ Prepare for and manage emergency situations such as fire, severe weather, or electrical blackouts.
- ✔ Recognize and react to serious injury or illness of guests.

Equally important as knowing good serving techniques and carrying out the duties of your job is using safe and sanitary work routines to protect your guests, coworkers, and yourself from accidents and foodborne illnesses. Being prepared for a quick response in any emergency, such as a fire, severe weather, or an electrical blackout, and knowing what to do for an injury or illness is important to ensure that everything within reason is done for the health and safety of you, your coworkers, and guests.

Safety: Preventing Accidents

Accidents are unpredictable, but careful attention to preventing accidents and preparation for quick response to accidents that do happen is important in the restaurant business. Every employee should be mindful of a safe environment.

Develop a routine with other servers for a safe restaurant operation. Have a well-stocked first-aid kit, including protective gloves, in case of an accident. Know the proper procedure for handling or working with cleaning substances. Report to management any safety hazards you see so that they can be corrected immediately. Use the following suggestions to make your restaurant a safer place to work or visit:

- Wear slip-resistant shoes with low heels.
- Pass other workers on the right in the aisles between tables.
- Look before pushing open a door so you do not bump into someone on the other side. Pull, do not push, carts through doorways.
- Keep work and serving areas clean and orderly. Keep aisles clear at all times. Beware of tripping over purses or briefcases that may be in an aisle (Figure 6-1).
- Keep exits clear and unlocked during operating hours for emergency evacuation.
- Let parents place and secure their child in a high chair or booster seat. The safety of the child is the responsibility of the parent.
- If kitchen floors are slippery, inform management. Managers may have to improve floor-cleaning procedures and/or add mats to slippery areas. Causes of slippery floors in the kitchen include spills, water on the floor from leaking pipes, slow drains, dishwashers, ice machines, or faulty walk-in freezer door seals, as well as unsafe employee work habits.
- Carefully follow manufacturers' instructions for preparing sanitizing solutions for wiping tables. Never mix cleaning materials. The Occupational Safety and Health Administration (OSHA) requires businesses to have **Material Safety Data Sheets (MSDS)** available, providing workers with the proper procedures for handling and working with chemical substances. Refer to these sheets for directions on how cleaning products are to be used and what steps need to be taken if the product is accidentally inhaled, ingested, or gets on skin or in eyes. Know the designated place where you can rinse the chemical from your skin or eyes to prevent injury.
- Immediately report or clean up any spilled food or beverage in the dining room. Put a chair over the spill while you get the proper supplies for cleaning it up. The most frequent guest mishaps in restaurants are slips and falls.

FIGURE 6-1 Obstacles in the Way of Serving: Always check for obstacles such as briefcases, parcels, or purses that may have been placed on the floor, in the way of serving, because they can cause accidents.
Photo by S. Dahmer

- The temperature of hot beverages should be regulated in the kitchen so that beverages are hot (approximately 160°F or 71.1°C) but never scalding. To prevent guests from burning their mouths, alert the guest that you are about to pour a hot beverage. Pour hot beverage refills only when guests request them or agree to a refill. If guests are seated close together, shield the guest with a clean, folded towel or napkin as you pour. Be sure the cup is on a table or tray when you pour a hot beverage into it; never pour a hot beverage while the guest is holding the cup.
- Stack dishes on trays so that they are well balanced. When picking up food in the kitchen, place full glasses in the center of the tray, and make sure handles

and spouts are well within the edge of the tray. Soup bowls and coffee cups should be placed directly on the tray so they do not spill on the saucers. When clearing tables, never stack glasses one inside the other. Stack dishes only to a reasonable height to avoid the hazardous juggling act that so often results in breakage and injury.

- Lift a stacked tray from a tray stand in the following manner: Stand close to the tray with your feet spread for balance. Bend your knees, but keep your back straight and not twisted. Place your left hand (nonserving hand) under the center of the tray. Grasp the edge of the tray firmly with your other hand. Breathe in to inflate your lungs, and then smoothly lift the tray to shoulder height (Figure 6-2). Bending your knees and lifting with your leg muscles prevents back strain.

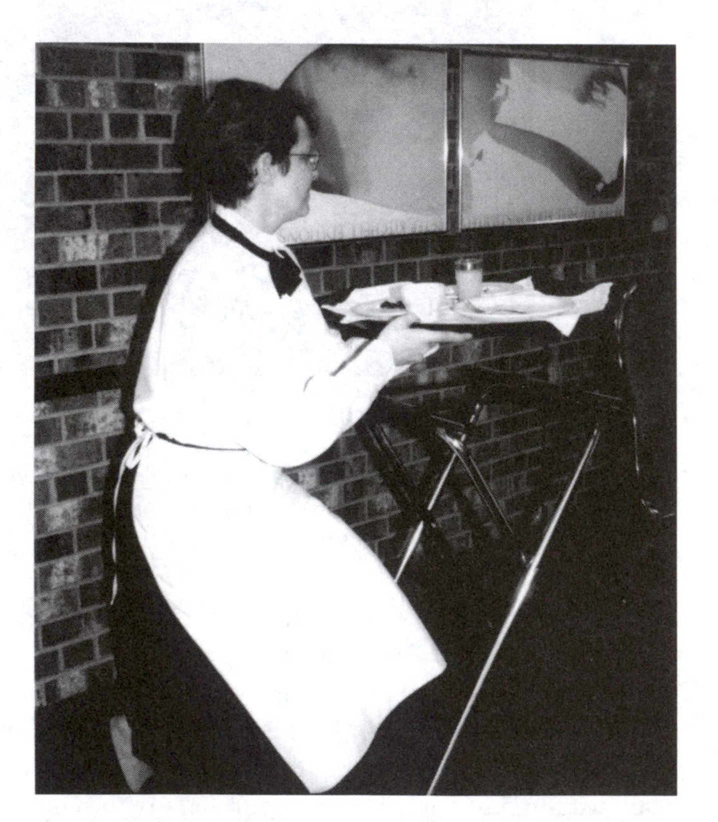

FIGURE 6-2 Server Lifting a Tray: Lift a heavy tray by bending your knees, keeping your back straight, and smoothly lifting the tray to shoulder height.
Courtesy of Hennepin Technical College, Eden Prairie, Minnesota; Photo by S. Dahmer

- Never lift trays of food or dirty dishes over the heads of guests.
- Never pick up several glasses in one hand by inserting your fingers into the glasses. If you do break a glass, use a broom and dustpan or a damp paper towel or cloth, not your bare hands, to pick up the pieces.
- If you do cut yourself or if anyone in the restaurant is injured or ill, exposing others to blood or other body fluids, the body fluid spills must be handled with a procedure called Universal Precautions. **Universal Precautions** reflect a standard by which all blood (and other body fluid) is treated as if potentially infected with HIV (human immunodeficiency virus) or HBV (hepatitis B virus). HIV and HBV are diseases that are communicable and difficult to treat or untreatable. Universal Precautions are guidelines that detail the cleaning up of potentially dangerous body fluids while wearing personal protective equipment (PPE), such as gloves and other protective clothing, and decontaminating surfaces and disposing of stained rags to ensure a zero risk of infection (Figure 6-3).

Food Safety and Sanitation Responsibilities

Food safety, or the safe handling of food, is an important public health priority in the United States. Every employee in a restaurant is responsible for using proper food-handling procedures, maintaining safe and sanitary food and eating conditions for guests.

Improper food and utensil handling can lead to contamination and foodborne illnesses. **Foodborne illnesses** are carried or transmitted to people from unsafe food. Each year, millions of people become infected by foodborne pathogens in food they eat, resulting in thousands of cases of foodborne illness and death.

The **Centers for Disease Control and Prevention (CDC)** is the government agency that strives to protect people's health and safety in the United States and around the world by providing reliable health information. While the food supply in the United States is one of the safest in the world, the CDC reports that 76 million illnesses, 325,000 hospitalizations, and 5,000 deaths are attributable to foodborne illness each year.

The **U.S. Food and Drug Administration (FDA) Food Code** is a model of sanitation rules and regulations that every state adopts and implements as its legislature requires. It provides requirements for safeguarding food and ensures that unadulterated and honestly presented food is offered to the consumer (guest). The Food Code defines a **foodborne disease outbreak** as "an incident in which two or more persons experience a similar illness after ingestion of a common food."

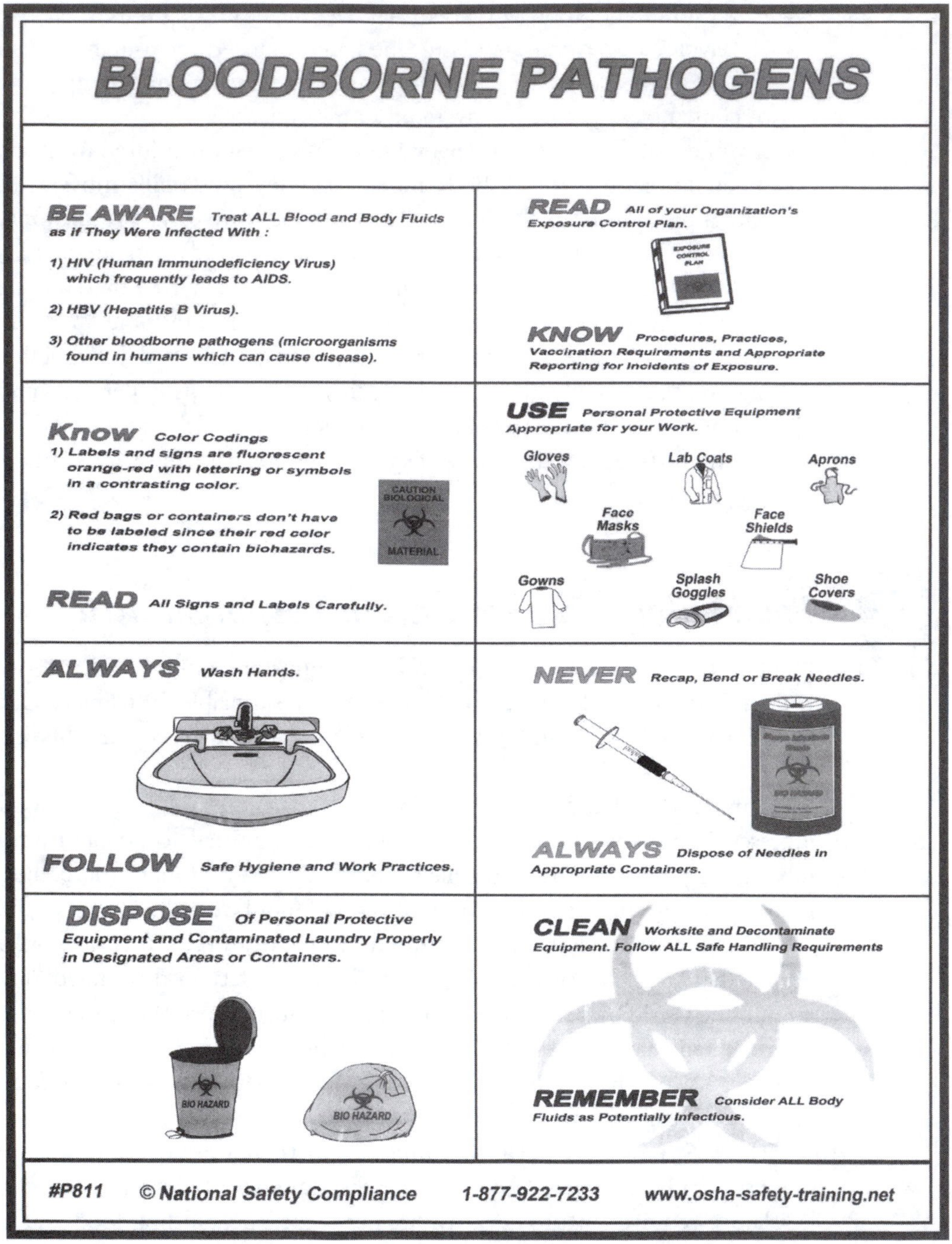

BLOODBORNE PATHOGENS

BE AWARE Treat ALL Blood and Body Fluids as if They Were Infected With :

1) HIV (Human Immunodeficiency Virus) which frequently leads to AIDS.

2) HBV (Hepatitis B Virus).

3) Other bloodborne pathogens (microorganisms found in humans which can cause disease).

READ All of your Organization's Exposure Control Plan.

EXPOSURE CONTROL PLAN

KNOW Procedures, Practices, Vaccination Requirements and Appropriate Reporting for Incidents of Exposure.

Know Color Codings

1) Labels and signs are fluorescent orange-red with lettering or symbols in a contrasting color.

2) Red bags or containers don't have to be labeled since their red color indicates they contain biohazards.

CAUTION BIOLOGICAL MATERIAL

READ All Signs and Labels Carefully.

USE Personal Protective Equipment Appropriate for your Work.

Gloves
Lab Coats
Aprons
Face Masks
Face Shields
Gowns
Splash Goggles
Shoe Covers

ALWAYS Wash Hands.

FOLLOW Safe Hygiene and Work Practices.

NEVER Recap, Bend or Break Needles.

BIO HAZARD

ALWAYS Dispose of Needles in Appropriate Containers.

DISPOSE Of Personal Protective Equipment and Contaminated Laundry Properly in Designated Areas or Containers.

BIO HAZARD
BIO HAZARD

CLEAN Worksite and Decontaminate Equipment. Follow ALL Safe Handling Requirements

REMEMBER Consider ALL Body Fluids as Potentially Infectious.

#P811 © National Safety Compliance 1-877-922-7233 www.osha-safety-training.net

FIGURE 6-3 Universal Precautions Chart: This Universal Precautions Chart provides guidelines for employees exposed to blood or other infectious materials to ensure zero risk of infection to the server. Compliments of National Safety Compliance, Inc.

Laboratory analysis must be done to confirm that the food is the source of the illness. The CDC has stated that, where reported, foodborne outbreaks were caused by mishandling of food.

The **Hazard Analysis Critical Control Point** (HACCP) is part of the Food Code. It is a food safety system that can be used to prevent outbreaks of foodborne illnesses through safe food handling. It covers identification of foods and procedures that are most likely to cause foodborne illnesses, builds in procedures that reduce risks of foodborne outbreaks, and establishes monitoring procedures to ensure food safety. Through HACCP, the FDA hopes to achieve uniform and effective standards of food safety for foodservice, retail stores, institutions, schools, and other retail-level establishments nationwide.

The following suggestions can help servers prevent food contamination in restaurants in which they work:

- Wear effective hair restraints to prevent hair from getting in the food or on food preparation surfaces.
- Make sure your uniform, apron, hands, forearms, and fingernails are clean to avoid the transfer of harmful bacteria to the food and utensils.
- Wash your hands after every chance of contamination, including before starting work and after using the restroom, handling money, or handling soiled utensils and equipment. Also wash your hands after coughing, sneezing, using a handkerchief or disposable tissue, using tobacco, eating, or drinking. Use proper cough etiquette. Keep your hands away from your hair, scalp, and face (Figure 6-4).
- Clean your hands and exposed portions of arms with a cleaning compound in a lavatory primarily designated for hand washing. Hands should be washed by vigorously rubbing together the surfaces of your lathered hands and arms for at least 20 seconds and thoroughly rinsing with clean water. Pay particular attention to areas beneath the fingernails and between the fingers. Rinse hands thoroughly under running water. Turn the faucet off using single-use paper towels. Dry your hands with paper towels or a warm-air hand dryer. Never use aprons or wiping cloths to dry hands.
- Do not work if you have an open wound, a cold, the flu, or any other communicable or infectious disease. Notify your supervisor so he or she can find a replacement. Cover any boil or infected wound with a dry, tight-fitting, sanitary bandage.
- Do not work if you have or have been exposed to noroviruses, hepatitis A, *Escherichia coli* (*E. coli*), *Salmonella typhi*, or *Shigella* until you have medical documentation that you are free of such an illness.

FIGURE 6-4 Wash Your Hands: To ensure sanitary eating conditions for guests, wash your hands properly before starting work and as often as necessary during work.

- Handle clean dishes by the rims, glasses by their bases, and silver by the handles to avoid contamination of food or lip-contact surfaces (Figure 6-5).
- Store tableware in a clean, dry place at least six inches above the floor and protected from flies, dust, splashes, and other contaminants.
- Wipe tables and counters with a cloth that has been stored in a sanitizing solution. Do not use wiping cloths for any other purpose. Use a second cloth from separate sanitizing solution to wipe chairs.
- Bring a clean utensil to replace one that has fallen on the floor.
- Keep dirty dishes completely separate from foods to be served to prevent contamination. Clear one course completely, removing it on a tray, before you bring the next course on another tray.
- Be aware that cold foods should be held at temperatures below 41°F (5°C), and hot foods should be held at temperatures above 140°F (60°C). Foods should be stirred on a regular basis during holding.
- Never mix new food with old food.
- Pour refills from a pitcher, wine bottle, or coffee server, avoiding contact with the guest's glass or cup. This will prevent cross contamination from one guest's glass or cup to another.
- Notify guests that clean plates are to be used each time they return to self-serve areas such as buffets and salad bars. A polite way to do this would be to collect soiled dishes and, at the same time, encourage guests to take another clean plate for refills.

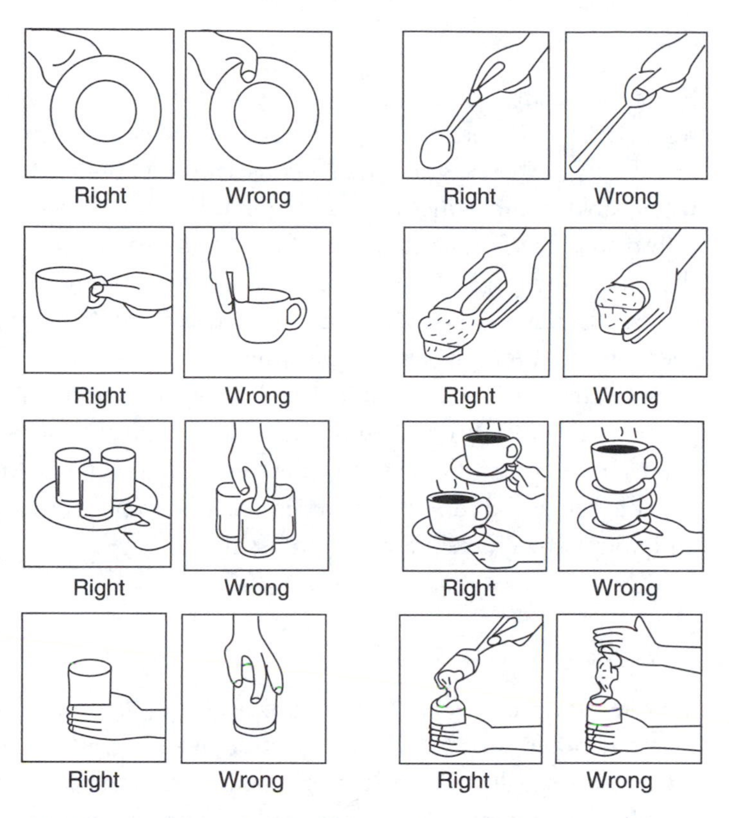

FIGURE 6-5 Handling Food and Tableware: To keep tableware sanitary, always handle dishes by the rims, glasses by their bases, and silver by the handles. Always use utensils, instead of your hands, to handle foods.
Courtesy of *Food Management*, March 2000

Emergency Procedures

Restaurant staff may have to manage a crisis at some time during their employment. Staff can be mindful of safety, but cannot prevent storms or, usually, fires or electrical blackouts, nor predict when someone in the restaurant will become injured or seriously ill. To be prepared for any incident, attend weekly meetings to review emergency procedures so they become second nature. The staff's main responsibility is to encourage everyone to remain calm, to control panic and confusion, to have a plan of action, and, it is hoped, to prevent serious consequences. The following information on specific types of crises will help in an emergency.

FIRE

Fire is an occupational hazard to any kitchen-based business, and fires occur in restaurants every year, endangering people and causing millions of dollars in property damage. A serving staff that is conscientious about safely handling open flames when cooking and using candles, keeping electrical equipment in good repair, carefully disposing of the contents of ashtrays into fireproof containers, and monitoring their own smoking habits is taking the first steps in fire prevention.

Because fires do occasionally occur despite staff precautions, be prepared to keep order and begin evacuation immediately. Alert the fire department either from the restaurant phone, if it can be done safely, or from an outside phone. Keep guests calm and get them out of the building quickly, following posted evacuation routes. Have a plan to cover all exits to make sure guests and coworkers leave and do not reenter the building. Reassemble at a preplanned gathering point outside, and notify firefighters if anyone is missing.

SEVERE WEATHER

Weather emergencies are frightening realities. We cannot control when or where severe weather, such as hurricanes, straight-line winds, earthquakes, snowstorms, or tornadoes, will strike, but we can take some precautions to minimize danger to people and property. Hurricanes and snowstorms are somewhat predictable and may entail management closing the restaurant to allow guests and staff to stay home. Of course, guests with reservations would have to be notified of the closure. Snowy days of a more minor nature and rainstorms would necessitate careful attention to slippery entrances and floors as guests enter the restaurant. Rain and wind may keep guests at their tables longer.

Be alert to changes in the weather by having one member of the staff listen to a local or National Weather Service radio station with information provided by the National Oceanic and Atmospheric Administration (NOAA). Severe storms, like tornadoes, require fast action on the part of the restaurant staff. Learn the warning signals used in your area—know the difference between a "watch" and a "warning." A **tornado watch** means conditions are favorable for tornadoes to develop. A **tornado warning** means that a funnel has actually been sighted in your area. Be prepared to act quickly in the event that you hear a tornado warning siren. If a warning is issued, help guests seek shelter immediately! Warn guests to stay inside and not to take cover in cars in the parking lot. Guests and staff should seek shelter in basements or inside rooms such as the kitchen away from large windows. Encourage everyone to lie low with their hands covering the back of their heads to reduce the possibility of neck injury. Stay sheltered until the storm is over.

Earthquakes can happen without warning. If an earthquake strikes, guests and staff should take cover under heavy tables, near an inner wall or door frame, at the inner corner of the room, or anyplace least likely to collapse and most likely to be a shield from falling debris. Do not use the elevator. Do not rush for exit doorways. Warn guests to stay away from glass windows and panels, chandeliers, furniture that may fall, or bricks that may come loose from fireplaces. Grab something to shield your head and face from falling debris and glass. Do not light a match or candle after an earthquake because of the possibility of a gas leak.

ELECTRICAL BLACKOUTS

Electrical failure may be temporary or long-lasting. If it is temporary, patience is your primary response, but if an electrical failure continues over a period of time, guests may be sitting in darkness at their tables and will need some lighting to continue dining or to move about safely. Most public buildings have emergency lighting that is triggered by an outage. If this is not the case, or if additional lighting is needed, distribute candles or battery-powered lights to tables as soon as possible to illuminate the restaurant.

INJURY OR SERIOUS ILLNESS

Injury and serious illness are often signaled by something unusual that catches your attention. Your senses—hearing, sight, and smell—may help you recognize an emergency. Unusual appearances or behavior that may signal an emergency are difficulty breathing; clutching the chest or throat; slurred, confused, or hesitant speech; unexplainable confusion or drowsiness; sweating for no apparent reason; or unusual skin color. Remain calm if a guest is seriously injured or becomes ill. Remember, your immediate responsibility is the guest's welfare. Knowing first-aid techniques may be valuable at this time. Know your state's Good Samaritan laws, which limit your liability if you help someone in a crisis.

It is advisable for the entire staff, or at least one person per shift, to be trained in first aid and personal safety, so they can react correctly in an emergency. **First aid** is defined as treatment for minor injuries or help until more complete treatment by medical personnel can be provided. Trained servers should know the technique of cardiopulmonary resuscitation (CPR) and First Aid for Conscious Choking Emergencies (see following section). Skills should be reviewed each year so responses become automatic and techniques are updated. A first-aid kit and nonlatex gloves for the treatment of injuries involving body fluids should always be on hand. Post the local emergency phone number or 9-1-1 near all restaurant phones.

Never move a guest who has been injured or is seriously ill unless there is an immediate danger or you need to move them in order to give care. Immediately enlist the help of the restaurant manager or other members of the serving team. Someone should be sent to call for medical help while others stay with the victim to comfort him or her and to start first aid, if necessary. Some of the serving team should relocate nearby guests to other tables as far away from the incident as possible.

If first aid is necessary, obtain the consent of the injured or ill person and explain what first-aid procedures will be done. Ask permission from a parent if a child is the victim. If the injured person is unable to communicate, you may give first aid because the law assumes the person would consent. For less serious injuries or illnesses, offer to call paramedics or a doctor, but let the victim decide.

One server should document the incident for insurance and liability purposes. Get names and phone numbers of people who witnessed the incident, and document what they saw happen.

Choking on Food

An emergency incident that may happen in restaurants is choking on food, such as a large piece of steak. The guest may turn blue from lack of oxygen, stop speaking, and desperately grasp at his or her throat. When a person choking on food is conscious and the airway is not completely blocked, you need not do anything other than remain at the scene and give support. Encourage the person to continue coughing. The hope is that the person will cough strongly enough to expel the food on his or her own.

If coughing does not dislodge the food and the throat is completely blocked, proceed with emergency procedures. Get permission to give care, and then call the paramedics (9-1-1 or the local emergency number). Open the airway immediately so the victim can breathe. Refer to the American Red Cross's **First Aid for Conscious Choking Emergencies**, which provides a list of recommended procedures to use when responding to a choking incident at a restaurant. If the adult or child cannot cough, speak, or breathe (choking), give five forceful back blows to the victim directly between the shoulder blades. In the event that the back blows do not dislodge the food, give five quick, upward abdominal thrusts as follows: Stand behind the victim and wrap your arms around the waist. Place the thumb side of your fist against the middle of the abdomen just above the navel. Grasp your fist with your other hand. Give five quick, upward thrusts to dislodge the food. Continue sets of back blows and abdominal thrusts until the person can breathe on his or her own. If the victim is unconscious, give CPR and look for an object in the mouth between giving compressions and breaths (Figure 6-6).

START HERE

1. CHECK

- CHECK THE SCENE for safety, then
- CHECK THE PERSON for consciousness.
  - Get permission to give care.
  - Tap shoulder and shout, "Are you okay?"

2. CALL

- IF NO RESPONSE, CALL 9-1-1 or have someone CALL 9-1-1 or local emergency number.
- IF YOU ARE ALONE AND CARING FOR A CHILD OR INFANT –
  - And you witnessed the child or infant suddenly collapse CALL 9-1-1.
  - If you did not see the child or infant suddenly collapse, give about 2 minutes of CARE, then CALL 9-1-1.

3. CARE

- OPEN THE AIRWAY (tilt the head back, lift the chin).
- CHECK FOR SIGNS OF LIFE (movement and breathing) for no more than 10 seconds.
- IF NO BREATHING, give 2 rescue breaths and begin CPR. For an adult or child use an AED if one is immediately available.
- IF BREATHING NORMALLY, roll onto one side while waiting for help to arrive.

CHOKING

- **If coughing, encourage the person to continue coughing.**
- **Send someone to CALL 9-1-1 or the local emergency number.**
- **Get permission to give care.**

INFANT CHOKING

If infant cannot cough, cry or breathe (choking)–

- Give **5** back blows

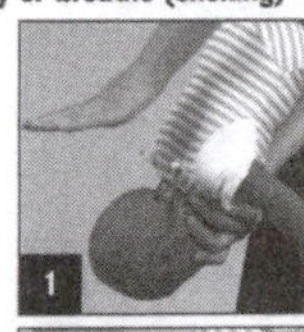

- If the object is not forced out– Give **5** chest thrusts

ADULT AND CHILD CHOKING

If adult or child cannot cough, speak or breathe (choking)–

- Give **5** back blows
- If the object is not forced out– Give **5** quick, upward abdominal thrusts

Continue sets of back blows and chest or abdominal thrusts until–

- Object is forced out.
- Person can breathe or cough forcefully.
- Person becomes unconscious. Perform CPR. Between giving 30 compressions and 2 breaths, look for an object and remove it if one is seen.

Local emergency number:

Emergency contact information:

Allergies:

Medical conditions:

This card is not a substitute for training. American Red Cross first aid, CPR and AED training can give you the skills and confidence to act in an emergency.

www.redcross.org

If no signs of life, give CPR

INFANT CPR

- Give cycles of **30** chest compressions and **2** rescue breaths
- If you are unable to give full CPR, give compressions only

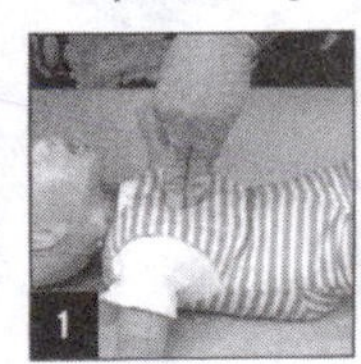

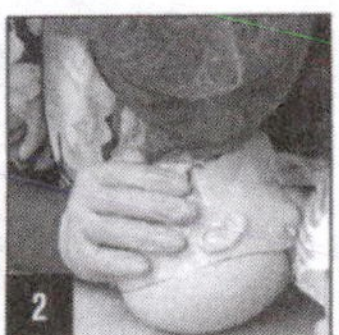

ADULT AND CHILD CPR

- Give cycles of **30** chest compressions and **2** rescue breaths
- If you are unable to give full CPR, give compressions only

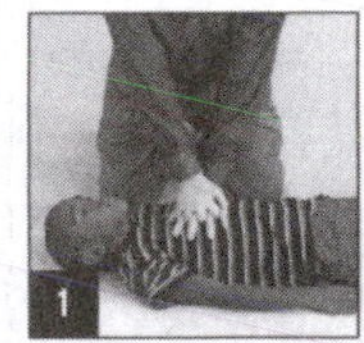

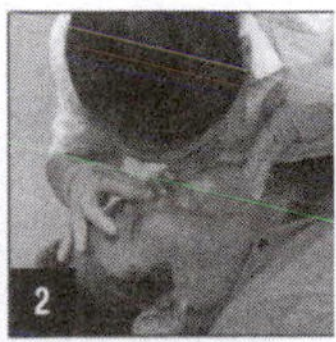

Continue CPR until–

- Scene becomes unsafe.
- You find an obvious sign of life.
- AED is ready to use (for adults and children over the age of 1)
- You are too exhausted to continue.
- Another trained responder arrives and takes over.

FIGURE 6-6 First Aid for Conscious Choking Emergencies: If coughing does not dislodge the obstruction causing guest to choke, call 9-1-1. While waiting for emergency paramedics, follow this procedure called First Aid for Conscious Choking Emergencies.

Be sure to watch the guest's belongings. Possessions such as purses, parcels, and coats should be kept in a safe place until the guest recovers.

In any emergency, you have an excellent chance to retain the respect of guests. You will fare well in a lawsuit if you have been conscientious about preventing accidents and have been sensitive to people involved in injuries and other emergencies.

KEY TERMS

- Material Safety Data Sheets (MSDS)
- Universal Precautions
- Foodborne illnesses
- Centers for Disease Control and Prevention (CDC)
- U.S. Food and Drug Administration (FDA) Food Code
- Foodborne disease outbreak
- Hazard Analysis Critical Control Point (HACCP)
- Tornado watch
- Tornado warning
- First aid
- First Aid for Conscious Choking Emergencies

REVIEW

1. Who is responsible for preventing accidents in a restaurant?
2. What is the safe way to pick up broken glass?
3. What does the term Universal Precautions refer to?
4. Why is washing your hands important for restaurant service?
5. When should you wash your hands while on the job?
6. List the sanitation responsibilities that involve personal grooming or hygiene.
7. What does a tornado warning siren indicate?
8. List all of the things you should do in the first 15 minutes after a guest suddenly sustains an injury or suffers a serious illness.
9. Why would it be helpful for a server to have CPR training?
10. How would you handle a situation where a guest bled from an injury sustained from a fall?
11. How would you handle a choking incident in a restaurant?

PROJECTS

1. Practice loading a tray safely with different combinations of tableware and foods.
2. Have a group discussion about ways to improve safety in a restaurant for the server, and then for the guest.
3. Place your unwashed fingers in a dish of agar-agar for a few seconds. Keep the dish in a warm place, and check it daily for bacterial growth. Observe the bacteria present on unwashed hands.
4. Have a health officer discuss with the trainees the sanitation laws of your city, county, and state as they apply to serving. In particular, discuss the laws relating to communicable and infectious diseases.

5. Write a one-page paper on foodborne illness discussing the number of people who become ill or die each year, the economic cost to society, and the procedures in place in restaurants to prevent an outbreak.
6. Invite a firefighter to demonstrate different kinds of fire extinguishers and how to determine which fires to use them on.
7. Post your local rescue squad telephone number next to the restaurant phone. Invite a local first responder to come to your training session and discuss first aid in situations of emergency.
8. Ask a representative of the American Red Cross to recommend a list of items to be stocked in a first-aid kit for a restaurant. Assemble the items, and explain the purpose of each at a staff meeting.
9. Attend an American Red Cross–sponsored training session. Learn and practice First Aid for Conscious Choking Emergencies.

CASE PROBLEM

Emergency Procedure

There is a severe weather warning in effect, and sirens are on in the city where your restaurant is located. You, as the server, are responsible for six tables, and you are the one whom guests would look to for questions and answers. The guests have heard the sirens. You have information about the severe weather. Answer the following questions:

- How do you go about informing the guests about the severe weather?
- Explain emergency plans for various forms of severe weather.
- List steps you can take to keep guests calm in severe weather situations.

Chapter Seven

Handling Service Using Technology

Objectives

After reading this chapter, you will be able to:

- ✔ Identify different types of restaurant point-of-service (POS) systems and technology.
- ✔ List the components that make up a restaurant POS system.
- ✔ Outline the procedure for taking orders with a restaurant POS system.
- ✔ Describe how orders get to and from the bar and kitchen.
- ✔ Explain how each transaction is settled at the end of service.
- ✔ Describe the procedure for closing a shift at the end of a day.
- ✔ List the advantages of restaurant technology.
- ✔ List the advantages and disadvantages of handheld order terminals.
- ✔ Define the features and advantages of a table management system.
- ✔ Define the features and advantages of a guest paging system.

National Restaurant Association (NRA) research shows that restaurant managers are increasingly relying on computers to remain competitive in today's marketplace. **Restaurant point-of-service (POS) systems** combine hardware and software to automate restaurant transactions and functions. The NRA reports that, "computers . . . can perform routine tasks that were once accomplished with paper and pencil and often do so more quickly, cheaply, and accurately." POS systems are always evolving to best serve the restaurant industry. Computer technology can help restaurant operators remain competitive in the midst of a shrinking labor pool and higher expenses.

This chapter provides current information about a restaurant POS system. It describes the usual components of the system, such as an office computer, server terminals with touch screen monitors, printers, cash drawers, magnetic card readers, and handheld order terminals. Handling the service electronically is discussed, with emphasis on taking the order, getting the order to the bar and kitchen, completing each transaction, and closing at the end of the day. The advantages of a restaurant POS system are also explained, as well as the advantages and disadvantages of handheld order terminals. Software programs for reservations/table management and the electronic device for guest paging, and the value of both, are explained.

Computers in Restaurants

A restaurant POS system may be as simple as a single computer stand-alone unit operated by a cashier or as complex as a multiple-computer system that links the host station, dining room(s), bar, kitchen, and office. Typically, restaurants have a terminal and printer in the dining room operated by the servers, a terminal and printer in the bar for the bartenders, and monitors or printers in the kitchen for the kitchen staff. POS systems in large, fine-dining establishments have multiple server stations, a host station, and an office computer for management (Figure 7-1). With POS systems, servers act as cashiers and complete the entire transaction for their guests.

Restaurant POS systems are set up so servers can create and store open checks, add a bar tab to the food order, communicate orders to the kitchen and bar, know immediately if a food item is sold out, send "fire" orders to the kitchen to start the next course, or signal a need for a utensil that has been dropped on the floor. With the POS system, guest checks can be totaled and tax added for the server. The system may be able to split a check between guests or split it evenly by the number of guests in the party. Guests' credit cards can be swiped for Internet-based credit card authorization. Gift cards can also be generated on the POS system. There could

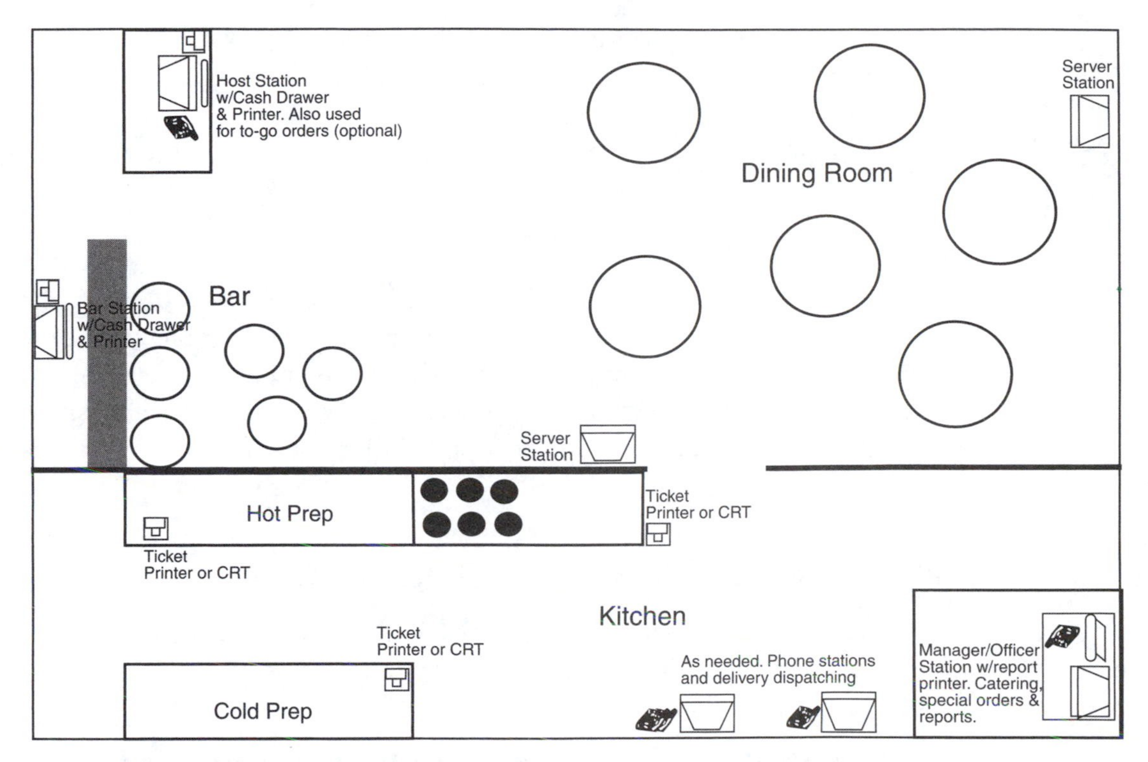

FIGURE 7-1 Diagram of a Restaurant POS System: A restaurant POS system has server terminals in the dining room and bar, and printers or monitors in the kitchen. The host station may have a terminal as part of the system to accommodate reservations and table management.
Courtesy of Rapidfire Software, Inc.

be a recipe lookup feature for the bartenders, and the system may be able to change pricing automatically for Happy Hour.

Establishments with a receptionist's desk can have a monitor with reservation and seating software to enable the maître d'hôtel or host to take reservations, manage seating, and keep tables full. Larger restaurants may also use a device for paging guests when a table is ready (see Restaurant Reservations and Table Management, and Guest Paging sections later in this chapter). Because of the efficiency of computers, servers can save time at the beginning of the meal when the orders are taken and submitted to the kitchen and at the end of the meal when the check is presented. Guests can reap the benefits of efficiency at the beginning and end of their dining experience, while enjoying their meal at a leisurely pace.

Restaurants of all sizes can benefit from using computers in their operations, although, according to the NRA, the most common users of computers are operations with large sales volumes and multiple establishments. Restaurants in all check-size categories will eventually use POS systems as computers become less expensive and more sophisticated.

Components of a POS System

OFFICE COMPUTER

A computer in the restaurant office controls all of the processes done by the restaurant POS system. **Software** (i.e., the application or instructions to run the computers) is installed by management on the office computer. Management reports, such as employee reports and inventory reports, can be generated for management on either the office computer or a dining room computer by use of a password.

SERVER TERMINAL WITH TOUCH SCREEN MONITOR

A **server terminal** is a touch screen monitor found in a convenient location in or near the dining room or bar (Figure 7-2). The touch screen allows the server to

FIGURE 7-2 Server Terminal: Servers and bartenders use terminals to enter orders into the POS system. Terminals usually have a touch screen, card reader, and printer. Photo by S. Dahmer

FIGURE 7-3 Touch Screen on Server Terminal: The server terminal may have a touch screen feature. The server merely touches the appropriate prompt on the screen to enter the order. Courtesy of MICROS® Systems, Inc.

enter information with the touch of a finger. The server merely touches the screen to enter the information and is guided through information input with visual prompts (Figure 7-3). Servers enter information, such as number of guests at a table and food and beverage orders, into the server terminal. The server terminal is programmed to send food and drink orders to the kitchen and bar to be filled. The brightness of the touch screen can be adjusted in dimly lit areas for easy readability. Information entered into the system is collected into reports that can be accessed by management from the office computer or any server terminal.

PRINTER

A **printer** imprints information on food and beverage orders, guest checks, receipts, and management reports. Drinks and food items, quantity, preparation, meal accompaniments, and other information can be clearly printed on paper in the kitchen and bar. Guest checks or tab receipts can be imprinted with the printer in the dining room, bar, or cashier's station. Management reports are stored for management use and are printed when accessed in the manager's office by use of a password.

CASH DRAWER

A **cash drawer** is divided by denomination into money compartments and placed close to, or attached to, the terminal. The addition of the drawer for making change, and a printer for printing receipts, converts the terminal to an electronic cash register (Figure 7-4). Cash drawers are commonly found at the cashier's station or at the bar.

MAGNETIC CARD READER

A **magnetic card reader** attached to the server terminal, bar terminal, or cashier's terminal reads numbers on a card pulled through a slot. Servers can log into the POS

FIGURE 7-4 Electronic Cash Register: An electronic cash register typically includes a cash drawer below the server terminal. Courtesy of MICROS® Systems, Inc.

system by swiping their authorizing card. They can also charge to a guest's credit card account by swiping the credit card. The magnetic card reader electronically charges the amount of the guest check to the guest's credit card account and verifies the transaction. Security is maintained because there are mandates by credit card companies that require high levels of security and encryption to prevent credit card fraud.

HANDHELD ORDER TERMINAL

A **handheld order terminal** is a small, portable, wireless POS system used by servers to take orders from guests at their tables (Figure 7-5). Because of its size, the handheld order terminal fits in a holster or case attached to the server's belt for immediate accessibility. The handheld order terminal either replaces the stationary server terminal or is fully integrated with it so that both can be used interchangeably in the dining room. The handheld order terminal is connected by radio transmission to the kitchen and bar. By use of a touch screen, a server can quickly send orders from the guest's table to the bar and kitchen, check on orders, or receive immediate feedback on item availability. The system can prompt the server for cooking temperatures and salad dressing choices and list items for suggestive selling. It can display photographs and access recipes of menu items so that the server can look up ingredients in foods. It can even access and update the accounts of frequent diners. Handhelds are built to withstand spills, high heat, and accidental bumps.

The handheld order terminal can be upgraded to include a compact magnetic credit card reader and receipt printer so that guests can pay at the table. Servers can handle all forms of payment, including cash, check, credit card, gift card, guest account, and even room charge. Servers can hand the device to guests so that they

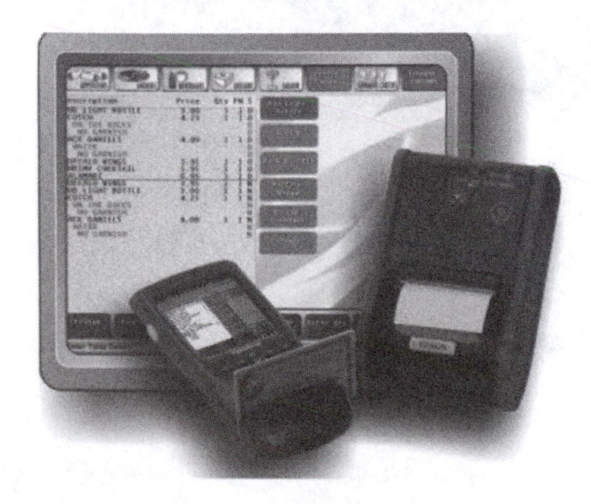

FIGURE 7-5 Handheld Order Terminal: Handheld order terminals can work with or replace the stationary server terminals in restaurants. Courtesy of Menusoft Systems, the developer of Digital Dining

can swipe their own cards at the end of the meal, and servers can get authorization and print a receipt without leaving the table. Split check, split item, and split payment can all be handled with the handheld unit.

Another version of the handheld order terminal is the **write-on handheld order terminal**, which works like a pencil and pad. Servers jot down an order by handwriting an abbreviation with a stylus, which quickly brings up the correct menu selection. Usually entering a couple of letters brings up all possible items from the menu, and the server can quickly tap on the appropriate item.

Handheld order terminals were initially recommended for busy bars, terraced dining rooms, alfresco dining, beach and poolside service, deck service on cruise ships, stadiums and arenas, and dinner theaters with short intermissions for serving drinks. Today, handheld order terminals are becoming popular in many traditional restaurants as well.

Taking Orders Using a POS System

In restaurants with server terminals, the server takes a guest's order at the table on a pad of paper. Having taken the complete order, the server proceeds to a server terminal. Each server is given a code number, authorizing card, or fingerprint identity to activate the server terminals in the dining room (Figure 7-6). The server

FIGURE 7-6 Authorizing Card: Each server is issued an authorizing card to activate the server terminals in the dining room. Photo by S. Dahmer

then enters the code or card or places an index finger on the finger pad to sign into the system. The server selects the table number and the number of guests to open an account for the party of guests. The system also automatically records the date and time of the order.

If a handheld terminal is issued to a server, an account is opened by the server at the guest's table without having to write the order first using pencil and paper.

The server then enters the order on the touch screen monitor by touching the screen prompts (Figure 7-7). As food and beverage items are entered, they are shown on the display for verification. Each entry asks for quantity, description, accompaniments, and preparation instructions.

FIGURE 7-7 Server Entering an Order: The server enters the order into the terminal in the dining room and does not have to walk to the bar or kitchen to place the order. Courtesy of MICROS® Systems, Inc.

The server may easily add to the guest's order at the same terminal or a different terminal in the same system. This is possible because the order remains open for each table and can be updated at any time. As each additional menu item is selected, it is added to the guest's total order.

Getting Orders to and from the Bar and Kitchen

Restaurant POS systems have monitors or printers in the bar, kitchen, special prep areas, and manager's office, depending on the system. As the server enters the guest's order at the server terminal or on the handheld terminal in the dining room, the information is immediately transmitted to the appropriate remote order monitor or printer (Figures 7-8 and 7-9). Food and beverage orders are filled by the chef and bartender without delay.

A restaurant with only a single kitchen may have a computer function identifying cold food items in blue ink and hot food items in red ink on the same paper from the kitchen printer. In a restaurant with a multiunit kitchen, one printer can be placed at the hot foods production station and another in the cold foods section. The POS system may be programmed to split the order and send each specific menu item to the appropriate kitchen printer.

A **drink runner** brings drinks from the bar, and an **expeditor** assembles and brings orders to the dining tables for the servers. This assistance helps facilitate efficient service and reduces kitchen traffic. In a restaurant without a drink runner or expeditor, the server is notified that the order is ready to be served.

Completing Each Transaction

Some restaurants have a cashier's station and a cashier to settle the guest's account. In other restaurants, the use of server terminals has eliminated the need for a cashier. Each server can act as cashier by using the terminal to authorize credit cards and by carrying enough money to make change for cash transactions.

When the guests have finished their meals, the server goes to the computer and totals all items ordered, automatically processing taxes and tips (if this is policy). The computer can even accommodate split checks, split items, and split payments if requested by the guest.

The itemized guest check is printed and presented to the guest for payment. Payment types include cash, check, credit card, gift card, guest accounts, and even room charge for hotel guests. If the guest pays in cash, the server takes the cash

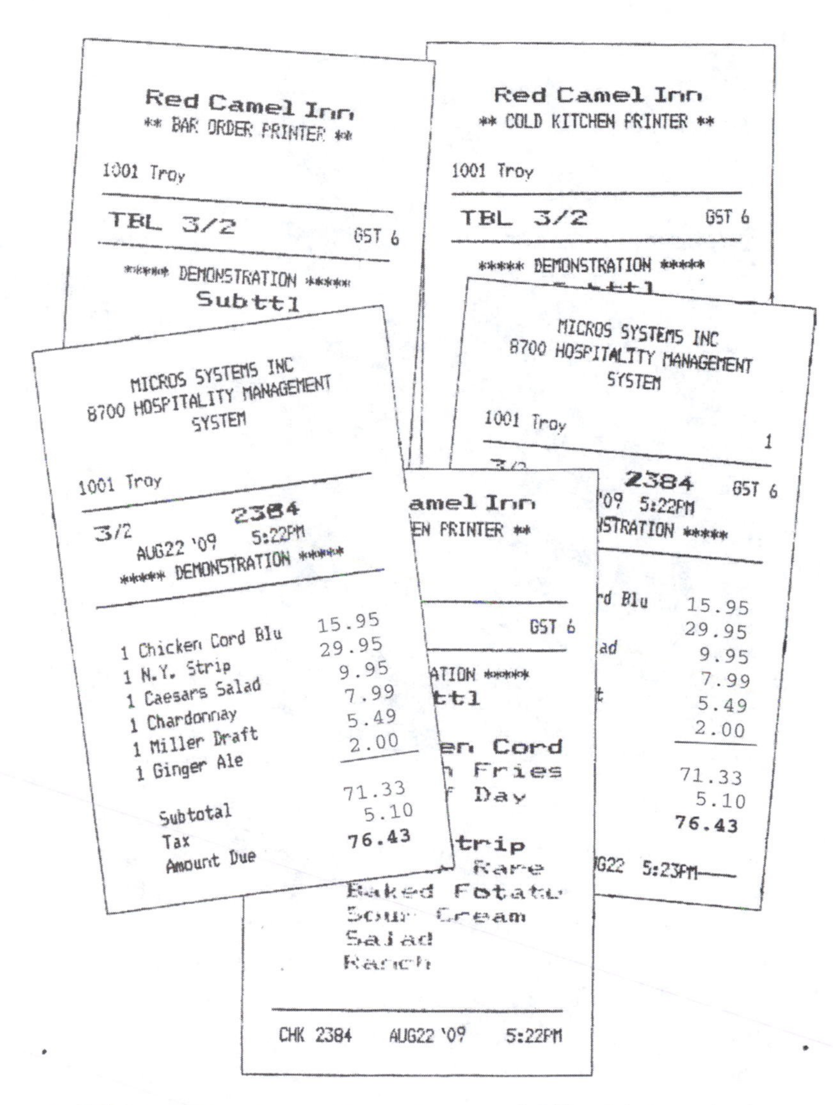

FIGURE 7-8 Bar, Kitchen, Dining Room, and Office Printouts: Each order entered in the POS system prints on the appropriate bar, kitchen, and dining room printers. The manager also has access to the order from the printer in the restaurant office. Courtesy of MICROS® Systems, Inc.

and the guest check to the terminal and indicates on the screen that the transaction is by cash. The computer can calculate the change for the server. The appropriate amount of change from the server's change pocket is taken back to the table and presented with the receipt. The server holds on to the cash payment and a copy of the paid guest check until the end of his or her shift.

FIGURE 7-9 Kitchen Monitor: With a POS system, servers send clear, legible timed orders to the chef on the kitchen monitor or printer. The chef then fills the orders without delay.
Courtesy of MICROS® Systems, Inc.

Payment by personal check is handled like cash, except identification must be provided (see The Guest Check and Payment in Chapter 5). In many cases, personal checks are no longer accepted at restaurants.

If the guest pays with a credit card, the server takes the credit card to the server terminal, swipes the card through the magnetic card reader (Figure 7-10), and waits for authorization. When authorization is given, a credit card guest check with lines for tip and total is printed. The server brings the credit card guest check, plus the credit card and a pen, back to the table to be signed by the guest (Figure 7-11). When settlement is complete with any form of payment, the guest's account is closed.

Closing at the End of the Day

If the restaurant has a cashier's station, the cashier keeps all receipts of settlements and balances them at the end of the day. If the restaurant has a POS system, the server keeps the cash, checks, and credit slips collected as payment during the shift. As the server enters each order into the computer, the amount of the item is automatically and immediately charged to that server's account. The computer

FIGURE 7-10 Swiping Credit Card: The server swipes the guest's credit card through the magnetic card reader at the server terminal.
Courtesy of PhotoDisc, Inc.

calculates all of the drinks and menu items sold, puts in appropriate taxes, and totals all of the guest checks for each server. At the end of the shift, the server cashes out completely, balancing the receipts and cash in his or her pocket with the guest check totals calculated by the computer. Tips are declared for the shift. Careful handling of all guest credit card information for privacy and security purposes is of utmost

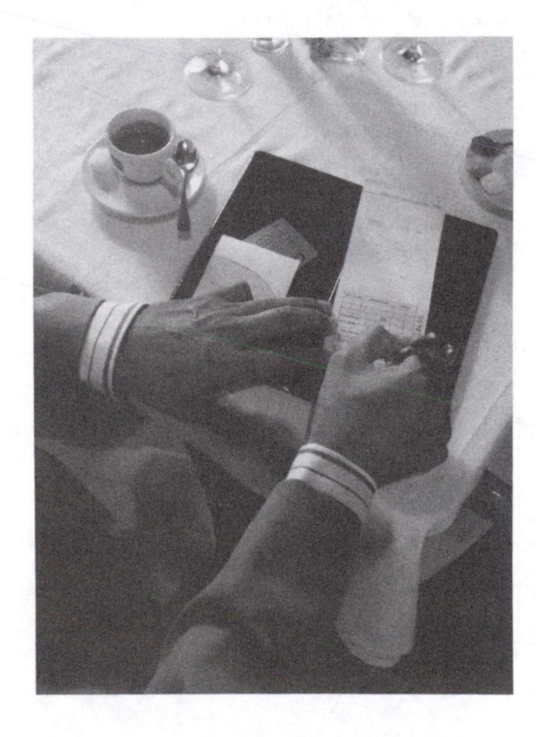

FIGURE 7-11 Guest Signing Credit Card Guest Check: A guest signs the credit slip generated by the computer.
Courtesy of PhotoDisc/Getty Images

importance. Reports like sales, payments, voids, profit, inventory, and table turns are generated to management.

Advantages of a POS System

Restaurants are using POS systems because they offer many advantages to the serving staff. The use of a POS system can:

- *Save steps in placing orders.* The average server walks miles a day on the job. Most of the walking is done taking orders to and picking up orders from the kitchen and bar. With a restaurant computer, the order is sent electronically, which cuts down on the number of steps the servers must take to the bar and kitchen.
- *Speed up service to the guest.* Placing orders electronically hastens the flow of information to the bartender and chef. Orders can be filled quickly, which speeds up service to the guest, increases the number of guests who can be served, and pleases them with prompt service. Reducing the time that it takes to serve guests at the beginning and end of the meal means tables are turned over quickly.
- *Increase server productivity.* Because servers spend less time walking back and forth and handling guest checks, they have more time to spend with their guests. They have time to make suggestions about the menu, answer questions, and serve correctly.
- *Ensure accurate food and beverage preparation.* With a POS system, servers send the bar and kitchen staff clear, legible, timed orders. There is no confusion over unusual abbreviations or illegible handwritten orders. The chef and bartender know which server ordered the food or drinks, the time the order was taken, for which table the food was ordered, the number of guests at the table, and exactly how the food or drinks are to be prepared.
- *Prevent confusion.* A hectic, fast-paced atmosphere is not always desirable in an elegant restaurant. Establishments with POS systems have less server traffic to the service bar and kitchen. A quieter, more relaxed mood can be maintained in the restaurant.
- *Prevent pricing errors on guest checks.* Restaurant computers automatically price menu items and drinks. The computer does all of the computations on the guest check, thereby eliminating errors in pricing, addition, tax calculation, discounts, and totals.
- *Produce itemized guest checks at server stations.* The restaurant computer automatically produces clear, accurate, concise, itemized guest checks at the end of

the meal. This feature eliminates the time-consuming delay as guests wait for servers to settle guest checks.

- *Generate management reports.* Advanced POS systems provide management with many kinds of management reports, as well as keep track of tips for tax purposes.

Advantages and Disadvantages of Handheld Order Terminals

The advantages of a computer POS system mentioned previously apply to handheld order terminals as well. In addition, the use of a handheld unit can:

- *Save more time at the beginning and end of service.* The server does not have to wait in line behind other servers to place or complete an order on a traditional POS server terminal. The server can use the extra time to better serve the guests. Servers can print checks immediately, process credit cards, and speed up the payment process at the table for expeditious check settlement. Quicker turnover of tables increases the opportunity for tips.
- *Provide on-the-spot information.* Knowing ingredients, accompaniments, wine lists, or sold-out items facilitates suggestive selling. When menu items are unavailable, the guest can be informed and make another selection immediately without having to be disappointed. After a table has been cleared, the server can communicate available table status to the host station in a timely manner.
- *Minimize errors in order entry.* The order goes to the kitchen and bar immediately without having to be reentered at the server terminal, where there might be the potential for error.
- *Ensure extra security for guests' credit cards.* The guests can feel secure in the fact that their credit cards never leave their sight. The cards are processed with the handheld magnetic card reader at the guests' tables.

Handheld order terminals are not without problems. The handheld unit can:

- *Have a small screen size with a complicated screen design.* The complicated format may not allow the server to give the guest their full attention while taking the order.
- *Be bulky.* The weight and thickness of the handheld unit may limit its portability.
- *Be expensive.* Converting a restaurant to a handheld system can be very expensive. A restaurateur must have one handheld unit for each server, whereas a

standard POS server terminal can be shared by four or five servers. Handhelds may also have a short battery life.

- *Have interference from surrounding equipment.* Microwaves and air conditioners can interfere with wireless signals and cause the handheld order terminals to malfunction. It can be difficult for wireless signals to travel through walls that contain steel.
- *Have security issues.* Managers fear that the network will drop transmissions or that data can be intercepted over the airwaves.

Restaurant Reservations and Table Management

Restaurant reservations and table management software has the functionality to track reservations as well as control seating and optimize server performance. It can centralize the dining reservation process, eliminate overbookings, or maximize table utilization with walk-in and waitlist functionality. Reservations can be entered and modified, and guest phone numbers, email, mailing addresses, and preferences can be captured (Figure 7-12). Table management software has a display of currently available and occupied tables (Figure 7-13). This system provides the host with the status of occupied tables and the approximate time when each table will again be open to seating. It can also show reserved and confirmed tables, tables to be joined by more guests in a party, tables needing to be cleaned, tables approaching or exceeding assigned departure times, and tables that are not to be utilized. From this information, waiting guests can be quoted realistic wait times, and servers can be given new parties of guests at a manageable pace in their stations.

Guests can even make their own reservations online over the restaurant's Web site. The host can then send the guests an e-mail confirming the reservation they made.

Advantages of Restaurant Reservation and Table Management Software

Reservation and table management software can:

- *Help the host seat parties of guests evenly throughout the restaurant.* New parties of guests can be seated on a rotational basis in server stations, because the host can see the overall activity on the screen.
- *Eliminate overbooking.* Table management software can keep the servers and kitchen staff from being overbooked and overburdened.

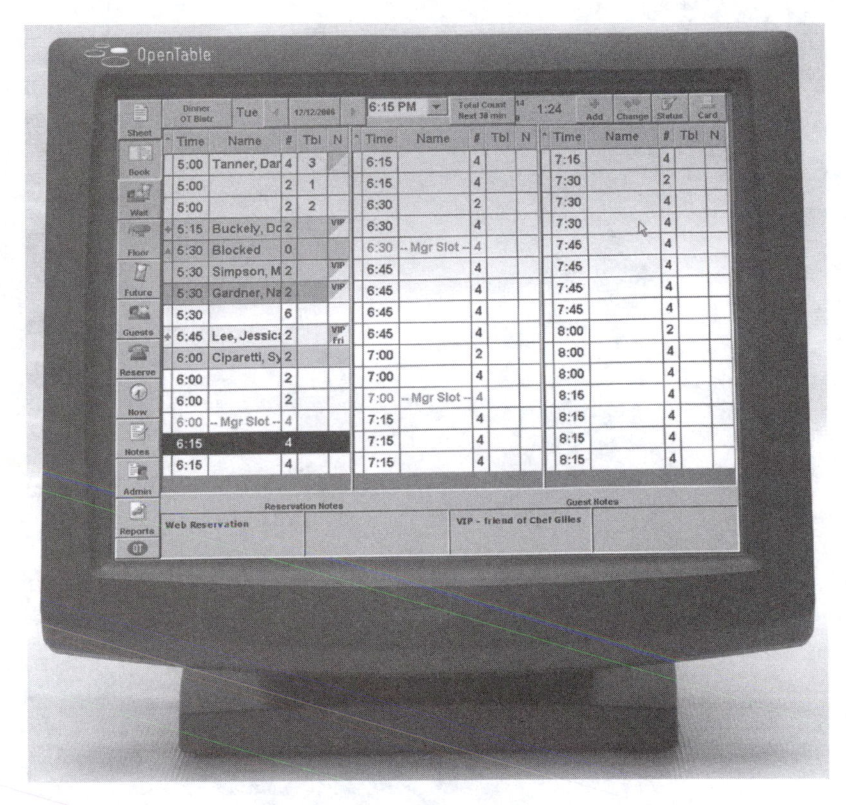

FIGURE 7-12 Reservation Management Software: Reservation management software will help track reservations in preparation for guests at an appointed time. Courtesy of OpenTable.com

- *Help the host give guests accurate waiting time quotes.* Guests can be informed how long it will be before they are seated. Guest walkouts will be reduced.
- *Provide the host with the guest's birthday, anniversary, table, and food preferences.* If this information has been kept on the computer, the host and server can use it to give more personal service to guests.
- *Increase table turns.* Efficiency provided by the seating software maximizes use of time at each table.

Guest Paging

A **guest paging device** is a wireless, remote guest call system that allows the host to signal the party of guests when their table is ready. The system includes coaster or paddle pagers, a charger, and a guest paging transmitter. Each party of guests is

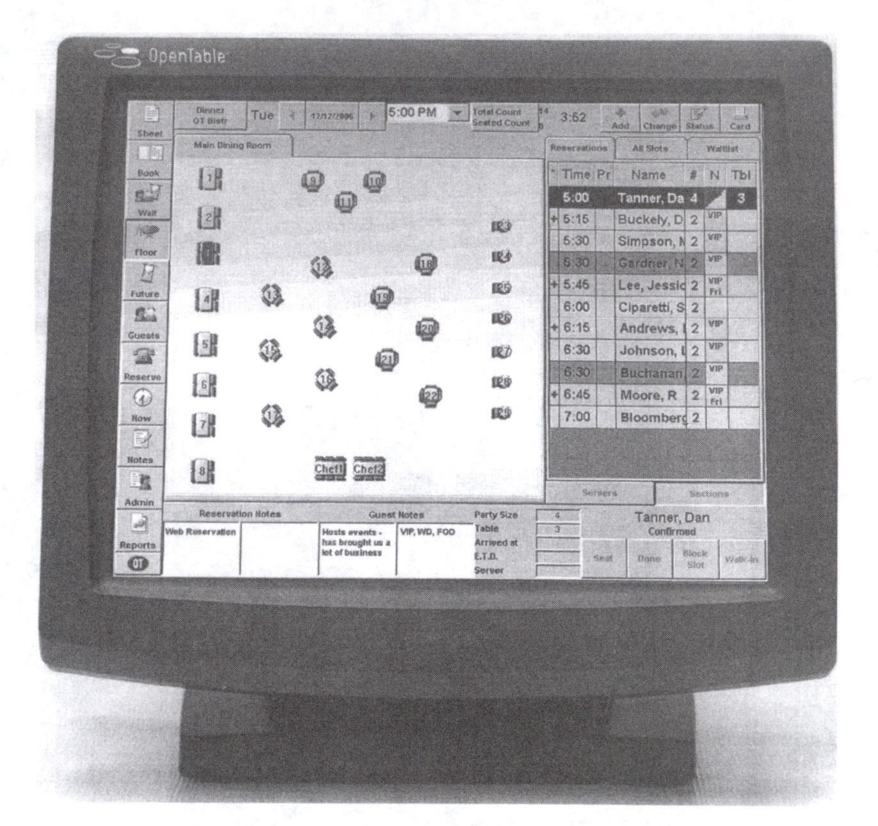

FIGURE 7-13 Seating Software: Seating software helps the maître d'hôtel or host quote wait times and seat guests in a more controlled manner. Servers receive guests in their stations on a rotational basis because the host can see the overall activity in the dining room.
Courtesy of OpenTable.com

given a pager that flashes, beeps, vibrates, or glows when signaled. Each pager has an adjustable range from ten feet to two miles (Figure 7-14).

Advantages of Guest Paging

Guest paging systems can:

- *Reduce guest response times to being called.* Guests are signaled quickly and quietly that their table is ready. Hosts or an assistant no longer need to search for guests throughout the entire establishment.

FIGURE 7-14 Coaster Pager: A coaster pager alerts the guest that his or her table is ready.
Courtesy of Long Range Systems, Inc.

- *Ensure that parties are seated promptly.* Tables no longer sit empty. Faster guest response means faster table turns, with greater revenues and profits for management and greater tips for servers.
- *Prevent crowds from forming at the host stand.* Guests no longer fear they will lose their table, so they do not hound the host for status reports every few minutes. A more inviting atmosphere is created.

KEY TERMS

Restaurant point-of-service (POS) systems
Software
Server terminal
Printer
Cash drawer
Magnetic card reader
Handheld order terminal
Write-on handheld order terminal
Drink runner
Expeditor
Restaurant Reservations and Table management software
Guest paging device

REVIEW

1. Describe a simple restaurant POS system. Describe a more complex restaurant POS system. What types of establishments might use simple or complex systems?
2. Define server terminal.
3. List and describe the components of a POS system in a restaurant.
4. What is a touch screen monitor?
5. What is the purpose of a magnetic card reader?
6. What types of information does the chef receive from the server on the kitchen printer?
7. What are handheld order terminals, and for what type of service situations might they be used?
8. How does a server add to a guest's order at a later time or from a different server terminal?
9. What kinds of computations can the POS system do on the guest checks?
10. What must be done at the computer terminal to settle and close each guest's account if payment is made in cash? If payment is by credit card?
11. What are the advantages of a POS system?
12. Why would a manager decide to spend the money for a handheld order system?
13. What are the advantages of a computer seating management program?
14. What are the advantages of a guest paging system?

PROJECTS

1. If you are a server in a restaurant with table service, or if you know a server who will help you, strap on a pedometer and keep track of the miles walked during an eight-hour shift. Do this three times and average the results. Discuss how a POS system, or the absence of one, would influence this distance.

2. Interview the manager of a restaurant with a POS system. Find out the complexity of the system he or she uses. Write a short paper discussing the step-by-step procedure the staff follows for using the computer, from taking the order to settling a guest's account. Include the backup plan or procedure that the restaurant uses when the computer is inoperative.
3. Compare the system of having a cashier with the system of having servers keep their own receipts. Discuss the advantages and disadvantages of each situation.
4. Talk to a restaurant host who uses reservation and table management software. What is his or her opinion of its usefulness?

CASE PROBLEM

Using the Computer

There is a new handheld POS system in place in the restaurant where you work. Some of the servers do not want to learn how to use the new system. You, as a server, have worked with the handheld computers, and the restaurant owner wants all servers to use the handheld system. Answer the following questions:

- What can you do as a server to encourage other servers to use the handheld computers?
- List the advantages of using the handheld POS system.
- What would be the advantages of knowing how to take orders with or without the handheld computer?

Chapter Eight

Wine and Bar Service

Objectives

Objectives: After reading this chapter, you will be able to:

- ✔ Understand the significance of serving wine, beer, and liquor in restaurants.
- ✔ Know your responsibility in regard to serving guests who are underage or intoxicated.
- ✔ Define wine and explain some of its characteristics.
- ✔ Distinguish among types of wine.
- ✔ Identify the temperatures at which wines should be served.
- ✔ Pair wines with foods they complement.
- ✔ Know how much wine to suggest and what type of glass to serve it in.
- ✔ Know how to open red, white, and sparkling wines.
- ✔ Define beer and list some types of beer.
- ✔ Identify types of beer glasses and describe how to pour beer.
- ✔ List common types of liquor.
- ✔ Know common types of liquor glasses.
- ✔ Identify many of the classic mixed drinks.
- ✔ Understand the importance of nonalcoholic drinks.

The service of alcoholic beverages is very important because restaurants are merchandising drinks to satisfy the preferences of guests and increase profits. The server's correct service of alcoholic beverages—wines, beers, and liquors—depends not only on knowing the procedure for serving but also on the knowledge of types of drinks, glassware, beverage temperature, and garnishes. A server should know which beverages complement particular foods on the menu in order to offer suggestions to guests.

Significance of Serving Wine, Beer, and Liquor

Many guests enjoy alcoholic beverages before, with, and/or after their meals when dining in a restaurant (Figure 8-1). Wines, beers, and liquors stimulate the appetite of the diner and are palatable companions to all types of foods. They not only make the meal taste better but also add a festive note to the pleasure of dining in a restaurant. Alcoholic beverages are among the most profitable moneymakers in the foodservice business. And because alcoholic beverages add to the guest's check, the server may enjoy a larger tip.

FIGURE 8-1 Guests Enjoying an Alcoholic Beverage: Many guests enjoy an alcoholic beverage with their meal. Alcoholic beverages add a festive note to the pleasure of dining in a restaurant.
Courtesy of PhotoDisc/Getty Images

A server should suggest drinks with dinner. Suggest a cocktail, appetizer wine, or beer prior to dinner, wine with dinner, or an after-dinner drink following the meal. Specifically asking whether the guest would like drinks and being especially knowledgeable about serving drinks add to your merchandising efforts (see Making Suggestions and Increasing Sales in Chapter 4).

The Concern about Serving Alcohol in Restaurants

Serving alcohol responsibly is of great importance to the server, the rest of the restaurant staff, the guests, and the community in which they live. Laws prohibit serving alcohol to underage or intoxicated guests and hold the server, bartender, manager, and owner of the restaurant personally liable for the guest's conduct and the consequences of excessive drinking even after the guest leaves the restaurant.

According to the Centers for Disease Control and Prevention (CDC), all 50 states have adopted a legal limit of 0.08 percent (or lower) blood alcohol concentration (BAC). If an individual's BAC is at or above the legal limit for a particular state, he or she is subject to arrest or revocation of driver's license if driving while intoxicated. Drivers under the age of 21 are not allowed to have any level of alcohol in their systems, because the consumption of alcohol under the age of 21 is illegal. A standard drink is a 12-ounce beer, 5-ounce glass of wine, or a drink with a $1\frac{1}{2}$ ounce shot of liquor.

Research by the CDC indicates that alcohol use slows reaction time and impairs judgment and coordination, all skills needed to drive a car safely. Some impairment begins well below the legal BAC limit for both males and females, and may even start after one drink, reports the CDC. The more alcohol consumed, the greater the impairment. It takes the average-sized woman approximately two drinks before reaching the 0.08 percent BAC limit, whereas an average-sized man can drink three.

The CDC states that 16,885 people in the United States died in 2005 in alcohol-related motor vehicle crashes, representing 39 percent of all traffic-related deaths. The CDC also reports that in 2006, alcohol-related motor vehicle crashes killed someone every 31 minutes and nonfatally injured someone every two minutes.

Servers and other restaurant staff have a responsibility to use caution when serving alcohol. The server's first responsibility is to only serve alcohol to guests of legal drinking age, and judging a guest's age can be very difficult. A server should verify the age of any guest who looks younger than 30 by asking for proof of age, such as a state-issued driver's license, a state-issued identification card, a military identification card, or a current passport.

A server must also identify the point at which a guest of legal age should discontinue drinking alcoholic beverages. The guest usually discontinues or slows the rate of alcohol consumption when dinner is served. Occasionally, however, a guest does not order dinner but consumes only alcohol or arrives at the restaurant inebriated. The server's responsibility is to handle the situation tactfully, with the welfare of the guest and the reputation of the restaurant in mind.

The server may be able to determine the degree of intoxication of a guest by monitoring the number of drinks consumed and identifying changes in behavior. Table 8-1, from the National Restaurant Association, identifies behavior associated with various levels of intoxication. This table is only a guide. From the table, you should try to identify guests who reach the yellow level. At this point, you can stall for time by serving drinks at a slower rate. Suggest coffee or nonalcoholic wines

TABLE 8-1 Guide to Recognizing the Stages of Intoxication

Green Level
(no noticeable behavior change)

Customer

- Is talkative
- Seems relaxed, comfortable, happy

Yellow Level
(relaxed inhibitions and impaired judgment)

Customer

- Becomes louder or more talkative
- Becomes more quiet
- Behaves in an overly friendly way
- Curses at the server's slow service
- Complains that drinks are getting weaker
- Insists on singing with the band

Red Level
(loss of motor coordination)

Customer

- Spills a drink
- Sways when walking
- Has slurred speech
- Asks for a double
- Is unable to pick up change
- Annoys or argues with other customers
- Becomes tearful or drowsy
- Has difficulty focusing
- Falls or stumbles

Reprinted with permission of the National Restaurant Association (NRA)

or drinks. Hand out menus for dinner. Suggest fatty and high-protein items to decrease the rate at which alcohol is absorbed into the bloodstream.

Alcohol service must be stopped before your guest reaches the red level. Tactfully inform the guest that alcoholic drinks will be discontinued. Show genuine concern for the guest and his or her safety getting home. Never be taken in by promises not to drive. Provide coffee and offer to call a taxi for an intoxicated guest. If the guest still insists on another alcoholic drink, turn the situation over to the manager on duty.

Some restaurants promote designated driver programs, in which one member of a party of guests who are consuming alcohol is encouraged to refrain from drinking and is designated as the driver for the group. The restaurant may provide free nonalcoholic beverages or food for the designated driver. Even with a designated driver, the rest of the guests in the party must not be overserved alcoholic beverages. A reputation for careful practices regarding alcohol service, a good alcohol policy, and a conscientious serving staff are a restaurant's best defense in a third-party liability lawsuit.

Wine Service

WINE MERCHANDISING

Wine is a beverage made from fermented grape juice and containing from 10 to 15 percent alcohol. Drinking wine is an important part of the American dining custom. Traditional wine service is given with much showmanship, and therefore the server needs to be fully acquainted with the procedure. Many restaurants have wine lists on the table or have the server present the wine list to the guest. Often, displays of wine bottles in the restaurant and the presence of wine glasses on the table subtly suggest wine as part of the meal. As a server, you should be ready to present and serve wine in the time-honored tradition at the guest's request.

CHARACTERISTICS OF WINE

Wines are identified by color, body, bouquet, taste, name, and vintage. As a server, you should be knowledgeable about wines and be able to distinguish one wine from another for your guests.

Wines are either white or red in accordance with age, grape variety, and the length of time the grape skins are left in the grape juice while it is fermenting to form wine. **White wines** are made by fermenting the grape juice without skins. **Pink**

wines (rosé and blush) are made by separating the skins soon after fermentation has begun. The skins remain in the tank to make **red wines**. White wines range in color from almost crystal clear to gold to amber, and red wines range from pink to red to crimson. **Rosé wines** are pink table wines and share characteristics of both red and white table wines, as do the blush wines. Rosé wines are usually blended wines, whereas **blush wines** are sweeter and are made from one grape variety.

Body describes the thickness or thinness of the wine. The body of a wine is determined by how it flows around the inside of a glass as it is swirled. A light wine flows quickly, and a full-bodied wine flows slowly. This light-bodied or full-bodied quality is also apparent on the tongue.

The **bouquet**, or aroma or fragrance, a wine emits as it is swirled is one of the most sensational qualities of a wine. For example, the bouquet may be fruity or flowery. Red wines need to rest or breathe for several minutes after they are opened to develop their bouquet.

The flavor of each wine ranges from very dry (not sweet) to very sweet. Complete fermentation produces a **dry wine**. A sweet wine is produced by interrupting the fermentation process. The dryness or sweetness characteristic usually determines when the wine is served during the meal (see Wines and Foods that Complement Each Other later in this chapter).

Wines may be named and identified on their labels by their varietal, generic, or proprietary names. Wines are often numbered on wine lists because their names are lengthy and/or difficult to pronounce. A description of names on labels follows:

- A wine with a **varietal name** has the name of the primary grape used in making the wine. California law states that at least 51 percent of the juice in a wine must come from the named variety of grape. Some wines contain up to 100 percent of the named grape. Examples of wines with varietal names are Pinot Noir and Chardonnay.
- Some wines are known by their **generic names**. These wines are named after the geographic region where the grapes are grown. Burgundy and Champagne are wines with generic names. American-made wines similar to the originals sometimes bear the generic name of the wine they resemble. For example, New York Champagne bears resemblance to French Champagne, and California Burgundy is similar to French Burgundy.
- Wines may have **proprietary names**, which are brand names adopted by the bottler for sales purposes. These names do not conform to other classifications.

Wines are produced in many countries. France, Italy, Spain, Germany, and Portugal are important wine producers, as are Argentina, South Africa, Romania, Australia, and others. The United States is also a major wine producer. Most U.S.

wines come from California, Oregon, Washington, and New York. American and European wines do not taste the same even when they are made from the same grapes. The difference in taste is caused by different climates, soils, and growing conditions.

Vintage refers to the wine from a grape crop or harvest of a good year, and many vintage wines carry the year of the harvest on their labels. In European countries, the vintage year statement is used to identify the years when the weather in the vineyard districts was sunny enough to fully ripen the grapes. In California, the long, rainless summers permit the grapes to mature every year, but the state's wines still reflect subtle changes in the weather from year to year. In order to state a vintage year on the label, the United States requires that 95 percent of the grapes in a wine be grown and fermented during that year. Wineries in other countries do not enforce this law, allowing for some very fine wines to be undated.

TYPES OF WINES

Wines are grouped into four basic types: table, sparkling, fortified, and aromatized.

1. **Table wines**. All unfortified white and red wines that normally accompany a meal are referred to as table wines. The alcohol content of table wines is 9.5 to 14 percent. They are created entirely by the natural fermentation of sugars in grape juice. Examples of table wines are Burgundy and Bordeaux.
2. **Sparkling wines**. Sparkling wines, often used for special occasions, contain 8 to 14 percent alcohol. Carbon dioxide is added to make these wines effervescent. Champagne is a well-known sparkling wine.
3. **Fortified wines**. Fortified wines are combined with brandy to increase the alcohol content to within 17 to 22 percent. Fortified wines can vary from dry to sweet. For example, a dry sherry can be used as an aperitif wine, and a sweet sherry or port can be served as a dessert wine.
4. **Aromatized wines**. A few wines are lightly fortified and flavored with herbs, spices, and peels. Aromatized wines, such as vermouth, are often served in combination with other alcoholic beverages as cocktails. Aromatized wines contain between 15 and 20 percent alcohol.

TEMPERATURE OF WINE

Red wine should be served at cool room temperature. White, rosé, blush, and sparkling wines are best served chilled (Table 8-2). The bouquet and taste of wines are at their best when wines are served at the correct temperature. Serving the wine at the correct temperature is the server's responsibility.

TABLE 8-2 Wine Serving Temperatures

Refrigerator Temperature	35 degrees
Ice Wines, Tokay, Berenauslese, and other dessert wines	43 degrees
Champagne, Riesling, Sauterne, Rosé	46 degrees
Chardonnay, Viognier	50 degrees
Beaujolais, Madeira	55 degrees
Chianti, Zinfandel	59 degrees
Cabernet, Red Burgundy, Bordeaux, Merlot, Shiraz, Pinot Noir, Port	63 degrees
Restaurant Temperature	70 degrees

Courtesy of LoveToKnow Corp.

Prior to service, all wines should be stored in a cool, dark location (Figure 8-2). An ideal wine storage temperature for a long period of time is approximately 55°F. Wines with corks should be stored horizontally so that the corks do not dry out and crumble into the bottle or allow air to enter and spoil the wine.

Before serving white, rosé, blush, and sparkling wines, chill them one or two hours in the refrigerator. To maintain a constant supply of chilled wines, some restaurants keep several bottles cold in the refrigerator at the bar. The server then gets the chilled bottle ordered by guests from the bartender and replaces it with a bottle of the same wine from storage, to be chilled for another time. Wines may be

FIGURE 8-2 Bottles Stored Horizontally: Bottles of wine should be stored horizontally in a cool, dark location to prevent the corks from drying out and to preserve the quality of the wine. Courtesy of PhotoDisc, Inc.

chilled in a refrigerator for up to two weeks; a longer period of chilling destroys the flavor. Wines that have been in the refrigerator longest should be used first. Never freeze or warm up a wine.

In some restaurants, a **wine cooler** or ice bucket may be used to maintain the proper temperature of wine at the table. Partly fill the bucket with ice and water so that the bottle slips in and out easily. Because this cooler is mainly for showmanship and maintenance of temperature, chilling wine in a refrigerator prior to service is still advisable.

Under normal conditions, red wines do not need chilling prior to service. To slightly lower the temperature, you may wish to chill the wine for 15 to 20 minutes in the refrigerator.

WINES AND FOODS THAT COMPLEMENT EACH OTHER

No hard-and-fast rules define which wine complements a certain food, although time, tradition, and common sense have given us certain wine and food associations. Therefore, any wine may be served with any food and be correct.

The server should serve the wine ordered by the guests and never criticize a guest's selection. When a guest asks your opinion, however, suggest a wine that is traditionally acceptable with the food ordered. Table 8-3 illustrates some well-accepted combinations of wines and foods that complement each other. Notice that dry, light wines are usually served at the beginning of a meal. As the meal progresses, medium-dry wines are served with the entrée. Delicate wines go well with lighter dishes, and full-bodied wines go well with robust foods (Figure 8-3). Only with dessert should sweet wines be served.

THE AMOUNT OF WINE TO ORDER

The normal portion of wine is four to five ounces per serving. Most bottles of wine are **fifths** (25.4 ounces or 750 ml). Wines in restaurants are also sold in half-bottles (tenths or splits) and by the glass. A good rule of thumb for ordering wine is a half bottle for two persons, a full bottle for three to five persons, and two bottles for six to ten persons. Knowing the approximate number and size of servings enables you to suggest the appropriate amount for your party of guests to order.

WINE GLASSES

A **wine glass** is traditionally a thin, stemmed, tulip-shaped glass, free of decoration and color. A serving of wine of only four to five ounces fills the glass approximately half full. Therefore, the bowl of the glass should be large enough to allow the guest

TABLE 8-3 Wines and Foods that Complement Each Other

Beef	Barbera, Barolo, Burgundy, Cabernet Sauvignon, Proprietary Red, Tempranillo, Pinot Noir
Cheese Dishes	Cabernet Sauvignon, Pinot Noir, Riesling . . . very ripe cheese hides the flavor of wine . . . serve fine wines with mild cheese
Chicken	Chablis, Chardonnay, Chenin Blanc, Riesling, Loire, Proprietary White, Sauvignon Blanc
Clams	Chablis, Chardonnay, Proprietary White
Crab	Chablis, Non-Vintage Champagne
Duck, Goose	Barbera, Cabernet Sauvignon, Merlot, Pinot Noir, Rhone, Blush, Zinfandel
Fish	Chardonnay, Chenin Blanc, Riesling, Sauvignon Blanc, Alsace White
Ham	Blush, Gewurztraminer, Zinfandel, Beaujolais
Hamburger	Cabernet Sauvignon, Merlot, Pinot Noir, Chianti, Zinfandel
Lamb	Chianti, Pinot Noir, Rhone, Zinfandel, Tempranillo
Liver	Beaujolais, Merlot, Rhone, Zinfandel
Lobster	Chardonnay, Gewurztraminer, Riesling
Omelettes	Chenin Blanc, Riesling, Proprietary White, Blush, Zinfandel
Oysters	Chablis, Chardonnay, Muscadet, Sauvignon Blanc
Pheasant	Cabernet Sauvignon, Tempranillo, Riesling, Pinot Noir, Rhone
Pizza	Everything
Pork	Zinfandel, Chenin Blanc, Alsace White
Quiche	Sauvignon Blanc, or a young Red according to ingredients
Scallops	Chablis, Chardonnay, Sauvignon Blanc
Shrimp	Sauvignon Blanc, Chardonnay, Gewurztraminer
Spaghetti	Barbera, Barolo, Chianti, Proprietary Red, Zinfandel
Turkey	Chenin Blanc, Riesling, Loire, Sauvignon Blanc
Veal	Cabernet Sauvignon, Pinot Noir, Zinfandel, Chardonnay, Riesling

Courtesy of the Pennsylvania Liquor Control Board

to swirl the wine and release the bouquet or aroma. The shape of the bowl should curve inward at the top to concentrate the aroma of the wine toward the nose. The stemmed feature allows the guest to hold a glass by the stem so the wine is not warmed by the heat of the hand.

A restaurant usually has several kinds of wine glasses suitable for serving different types of wine. Technically, there can be a wine glass size and shape for nearly every type of wine, but many restaurants use just a few main types of wine glasses as follows: appetizer and dessert wine glasses (4–5 oz.), tulip-shaped white wine glasses (7–9 oz.), red wine glasses with larger, more rounded bowls (9–12 oz.), and tall, thin flutes (7–8 oz.) for sparkling wines. Some restaurants may use only one large all-purpose eight-ounce wine glass suitable for all types of wine.

Because white wine is served chilled, a smaller serving is poured, allowing the remainder to stay in the ice bucket. Champagne was traditionally served in a saucer-shaped stemmed glass, but today a narrow tulip-shaped glass is preferred because it allows the wine to generate bubbles longer. With each new type of wine ordered,

FIGURE 8-3 Example of Wine and Food Pairing: Pair a wine with a food it will complement, such as this light-bodied red wine with duck. Photo by S. Dahmer

a clean, appropriate glass should be provided. Some traditional wine glasses used in restaurants are shown in Figure 8-4.

WINE LISTS AND WINE CHARTS

A **wine list** is a wine menu. A restaurateur works closely with a local wine distributor to place on the list dependable domestic and foreign wines that complement the food menu. A balanced wine list is made up of red, white, rosé, and sparkling wines.

FIGURE 8-4 Traditional Wine Glasses: Traditional wine glasses frequently used are the all-purpose wine, champagne saucer, champagne tulip, and sherry glasses. Today, a champagne tulip glass is preferred for champagne because it allows the wine to generate bubbles longer.

A good wine list is brief and describes the characteristics of the wines. It is generally kept simple, because the average guest recognizes only a few of the more popular wines. A wine list may include several expensive wines, but the bottles of wine on the wine list are generally priced at no more than the average price of a dinner. Your restaurant should have plenty of copies of the wine list on hand. These lists should be located in a convenient place and kept in good condition; messy, untidy copies with written-in changes should not be used.

A good server reads the complete wine list and learns how to pronounce the names of the wines in order to communicate them to the guests. Some restaurants have a **wine chart** carried by servers that describes available wines and pairs them with menu items. Most charts include the wine list number, name, phonetic pronunciation, year, price, type, origin, serving temperature, characteristics, and foods the wines complement. A good wine chart shows whether the wine is light or full-bodied, sweet or dry.

TAKING THE WINE ORDER

After becoming familiar with the wine list, you will feel comfortable taking the wine order. Bring out the wine lists with the menus and distribute them around the table. When only one wine list is available per table, open the wine list and hand it to the host of the table.

Take the wine order after you take the food order. You may sense a feeling of unfamiliarity with the wine list, and in this case you might suggest, "Our California Riesling is excellent with the broiled trout you ordered," or "We have a very good house rosé that will complement everyone's meal." Thank the guests, collect the wine lists with the menus, and continue with the service.

Serve the wine according to the time given to you by the host. If guests are enjoying cocktails first and have not designated when to serve the wine, open and serve wine with the entrée.

PROCEDURE FOR OPENING AND SERVING WINE

Serve wine in the proper manner, with a great deal of showmanship. Guests will get the greatest enjoyment from wine service as follows.

Opening Red Wines

Obtain a bottle of red wine of the right temperature and carefully carry it to the table. From the right side, present the bottle to the host by showing the label

FIGURE 8-5 Presenting the Wine to the Guest: A bottle of wine should be carefully carried to the table and presented to the host so the label can be read. Courtesy of Digital Vision at Getty Images®

(Figure 8-5). State the name of the wine, such as "The Burgundy you ordered, sir (or ma'am)." Wait for approval from the guest, then place the bottle of wine on the table to be opened.

The correct wine glasses are brought next so that the guests may anticipate the wine service. Set each wine glass on the table to the right of and slightly below the water glass. When serving several wines, either place the glasses in a line to the right of the water glass or place one wine glass below the other (Figure 8-6).

Handle the bottle carefully, so any sediment in it is not stirred up. Cut around the lower lip of the bottle with a sharp knife and remove the foil. Wipe away any mold that has formed near the cork. Press on the cork slightly to break the seal between the cork and the bottle. Using a **waiter's corkscrew** (Figure 8-7), insert the spiral screw into the cork and twist it until the corkscrew is almost completely inside the cork with only one turn showing. Place the lever on the edge of the wine

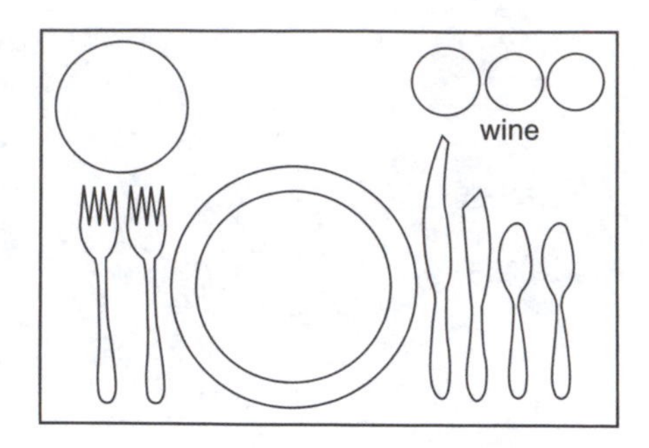

FIGURE 8-6 Ways of Placing Wine Glasses in the Cover: Either way of placing wine glasses in the cover shown here is proper when more than one kind of wine is to be served with the meal.

bottle and, using the handle, remove the cork from the bottle (Figure 8-8). The complete method of opening the bottle of wine is shown in Figure 8-9.

Smell the cork to detect any musty, vinegary, or sulfuric odor that indicates the wine is not good. If the cork has an off-smell, replace the bottle with another at once. If the wine is good, remove the cork from the corkscrew and place it on the

FIGURE 8-7 Waiter's Corkscrew: While there are many varieties of corkscrews, the waiter's corkscrew is especially popular. Photo by S. Dahmer

FIGURE 8-8 Server Opening a Bottle of Wine: When opening a bottle of wine, place the lever on the edge of the wine bottle and use the handle to remove the cork. Courtesy of PhotoDisc/Getty Images

table to the right of the host's glass. Wipe the lip of the bottle. Place the bottle of wine in the center of the table to allow the bouquet to develop.

Pour the wine for the guests when the entrée is served. Pour a sample of one or two ounces of wine into the host's glass (Figure 8-10). Once the wine has been tasted and approved, fill the other glasses at the table. Leaving each glass on the table, fill the glasses one-half to two-thirds full (a four- or five-ounce serving), twisting the bottle a quarter turn as you finish pouring to prevent dripping. Pour the wine for the woman at the host's right and continue counterclockwise, serving all of the women in the party. Then serve the men, serving the host last. Never completely empty a bottle, because you may pour out the sediment at the bottom. If the proper amount of wine has been ordered, there will be enough wine for each guest. Place the remaining wine on the table to the right of the host. Be available to refill glasses as needed. You may wish to suggest another bottle.

Opening White, Blush, and Rosé Wines

When a white, blush, or rosé wine is ordered, bring the chilled bottle to the table in a bucket of crushed ice and water and place it on a stand to the right of the host. Present the bottle by showing him or her the label and wait for approval. Then place the chilled wine back in the ice bucket.

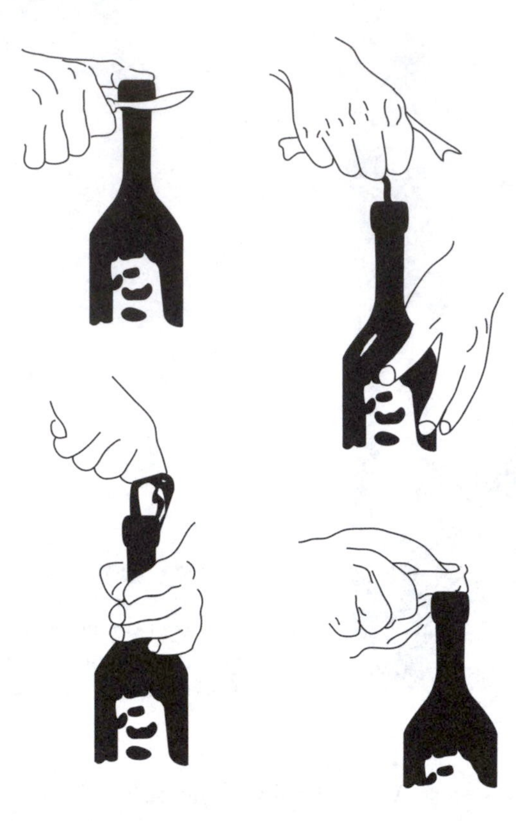

FIGURE 8-9 Opening a Bottle of Table Wine: When opening a bottle of table wine, cut away the foil and wipe away the mold, remove the cork with a corkscrew, and wipe the lip of the bottle again before pouring from it.

Bring the wine glasses next. Set each wine glass on the table to the right of and slightly below the water glass. When serving several wines, either place the glasses in a line to the right of the water glass or place one wine glass below the other, as shown in Figure 8-6.

Open the wine in the ice bucket to keep it chilled. Cut around the lower lip of the bottle with the knife, and remove the foil covering the top of the cork. Wipe away any mold that has formed near the cork. Press on the cork slightly to break the seal between the cork and the bottle. Using the corkscrew, insert the spiral screw into the cork and twist it until the corkscrew is almost completely inside the cork with only one turn showing. Anchor the lever on the lip of the bottle, and using the handle of the corkscrew, remove the cork from the bottle.

Smell the cork to be sure the wine is good. Replace a wine with a musty, vinegary, or sulfuric smell with another bottle. If the wine is good, remove the cork from the corkscrew and place it on the table to the right of the host's glass. Wipe the lip of the bottle with a clean napkin. Pour a sample of the wine for the host and wait for approval.

FIGURE 8-10 Pouring a Sample Taste for the Host: When serving wine, first pour a sample taste for the host and wait for his or her approval. Courtesy of Digital Vision at Getty Images®

Pour a white, blush, or rosé wine as soon as it is opened and approved by the host. Wrap the chilled bottle in a clean napkin to maintain its temperature. Allow the label to show. Pour glasses one-half to two-thirds full, and twist the bottle as you finish pouring to prevent dripping. Pour the wine for the woman at the host's right, and continue counterclockwise to serve all of the women in the party. Then serve the men, serving the host last.

Place the bottle with the remaining wine back into the ice bucket so that it remains chilled. When removing the bottle from the ice and water to refill glasses, wipe the water from the outside so you do not drip water on a guest or on the table.

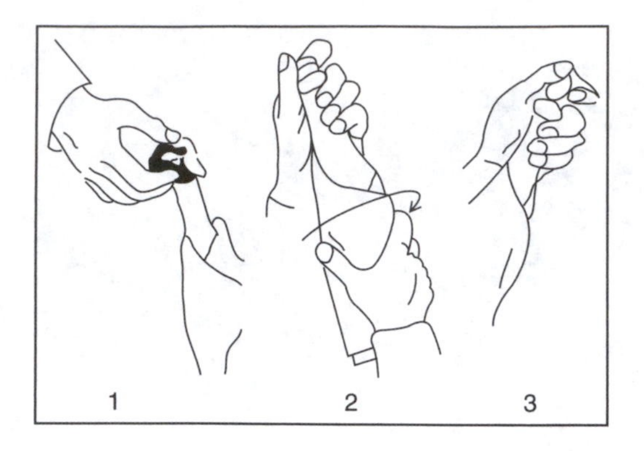

FIGURE 8-11 Opening a Bottle of Sparkling Wine: Open a bottle of sparkling wine by first removing the foil and wire. Then grasp the cork and tilt the bottle away from the guest. While holding the cork firmly, twist the bottle to allow the internal pressure to push the cork out.

Opening Sparkling Wines

As with white, blush, or rosé wines, keep the chilled bottle of sparkling wine in an ice bucket at the host's right. Bring wine glasses next, as described previously. To open the wine, wrap the bottle in a clean cloth napkin to maintain its chilled temperature and protect your hands from possible glass breakage. Cut the foil below the wire with the knife and remove it. Untwist the wire with your right hand while holding your left thumb on top of the cork. Remove the wire. Continuing to hold in the cork with your left thumb, tilt the bottle at a 45-degree angle and point it away from the guests. Firmly hold the cork with your left hand, and twist the bottle with your right hand. Let pressure escape slowly. Allow the pressure in the bottle to force the cork out gently without a pop or fizz (Figure 8-11). Remove the napkin from the bottle, and serve the host a taste. When given approval, serve the rest of the guests as you would with table wines.

Beer Service

TYPES OF BEER

Beer is a term referring to a brewed alcoholic beverage made from fermented barley malt, hops, yeast, and water with an alcoholic strength of 2 to 6 percent. In the United States, most of the beer consumed is a **lager beer**, a generic term for a pale, aged, effervescent brew introduced from Germany during the middle of the 19th century. In addition to malt, other grains such as corn and rice are frequently used to give lager its light body. All lager beers are aged by storing them for several months before putting them into bottles, cans, or kegs. Lagers should be served chilled (48°F/9°C).

Ale is another generic form of beer that differs from lager beer in that it has a different proportion of beer ingredients and is fermented at a higher temperature. These differences in brewing make ale a heavier beer with less carbonation. Ale has a more pronounced flavor of hops and higher alcohol content. Serve ale at a warmer temperature (54–56°F/12–13°C).

Types of beer have certain characteristics. For instance, **pilsner** is a lager beer with a pale, golden color, a strong hops flavor, and an alcohol content of 4 to 5 percent. **Munchner** is a beer made in Munich, Germany. It is deep brown in color and slightly sweet and has a strong malt flavor. **Weisse**, a white beer, is a German beer made from wheat. It is light and very refreshing. **Malt liquors** are lager beers with an alcohol content of 4 to 8 percent. **Light beer** is a lager specially brewed to reduce the number of carbohydrates and calories. Once a year, American and European brewers make a **Bock beer** and offer it in the spring. Bock is a dark beer with more body than usual and the added pleasant bitterness that comes from hops.

TYPES OF BEER GLASSES

Beer may be served in a mug, pilsner, goblet, schooner, tumbler, shell, stein, or hourglass tumbler. Pictured in Figure 8-12 are a **footed pilsner**—a tall, narrow, seven-ounce glass with a short stem—and a **stein**—a heavy glass with a handle. Beer glasses must be spotlessly clean, with no greasy film, to prevent the beer in them from losing carbonation.

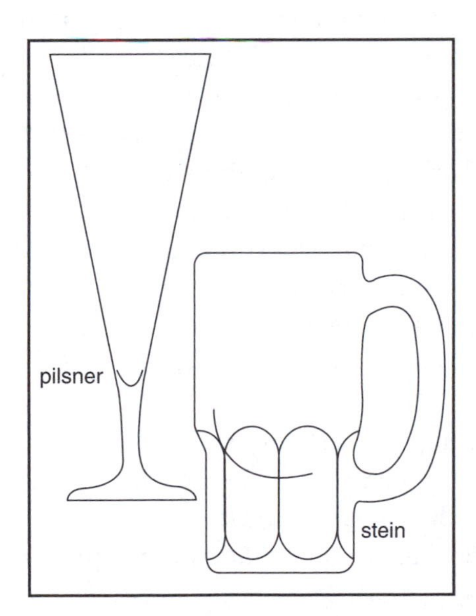

FIGURE 8-12 Two Beer Glasses: A footed pilsner and a stein are two of the many glasses used to serve beer.

PROCEDURE FOR SERVING BEER

Beer goes well with almost any food served in a restaurant except sweets. Beer may be served before the meal with the appetizer, during the meal, and as a beverage any time.

Pour the beer for the guest. Hold the glass at a 45-degree angle and begin to pour to release some carbon dioxide. At the halfway point, bring the glass upright and continue to pour into the center of the glass to create the perfect foam head. The foam head should be about one inch in depth and come to the top of the lip without spilling over. The head is very important because it releases the beer's aroma and flavor and is a nice presentation. Place the beer on the table to the right and below the water glass. Most of the rules for serving beverages (see Chapter 5, Serving the Meal) apply to serving beer.

Liquor Service

TYPES OF LIQUOR

Cocktails, mixed drinks, and straight drinks are made from brandy, whiskey, vodka, gin, rum, cordials, and other liquors or spirits. As a server, you should be familiar with the following liquors or spirits, which are the most frequently ordered:

1. **Brandy** is liquor made by the distillation of wine or a fermented fruit mash. The word brandy by itself refers to the liquor that is made from grape wine. Kirsch (cherry brandy), applejack (apple brandy), and apricot brandy are examples of fruit brandy. A well-known brandy is cognac, which is distilled in the region near the city of Cognac, France.
2. **Whiskey** is liquor distilled from fermented grain mash and aged in wooden barrels. The grain, yeast, water, and blending of whiskey have a lot to do with the flavor and lightness of the spirit. Most restaurant bars stock straight, blended, and bonded whiskeys that vary in age, alcoholic content, and flavor characteristics. Bourbon is a well-known straight whiskey. American, Canadian, Scotch, and Irish whiskey are the main whiskeys sold in the United States.
3. **Vodka** is usually distilled from grain mash but is not aged, as are many other spirits. It is colorless and has no aroma or flavor. Because of its neutral character, it is usually served blended with fruit juices, spirits, or wines.
4. **Gin** is redistilled liquor employing the juniper berry as the principal flavoring agent. Other flavoring agents are roots, herbs, peels, and other berries. Gin is a dry spirit used in many popular drinks.

5. **Rum** is a distilled beverage made from sugar cane by-products. The majority of rum production is in and around the Caribbean and South America. Rums range from the very dark Jamaican rums to the light, delicately flavored rums from Cuba and Puerto Rico. Rums are used for cooking and in many fruity drinks.
6. Other liquors include **aquavit**, a Scandinavian distilled beverage made exactly the same way as gin but with a caraway flavor, and **tequila**, a Mexican spirit distilled from the juice of the agave plant.
7. **Cordials** are sweet, colorful drinks that must contain at least 2 percent sugar. In Europe, cordials are called liqueurs. They are made by various processes that allow brandy to absorb the coloring, flavor, and aroma of fruits, leaves, and peels. Well-known cordials include crème de cacao, Benedictine, Chartreuse, Drambuie, crème de menthe, and Triple Sec.

TEMPERATURE OF DRINKS

Almost all cocktails, mixed drinks, and straight drinks must be served very cold. In some restaurants, the glasses are chilled to maintain the cold temperature of the drinks served in them.

LIQUOR GLASSWARE

Glasses vary in size and style, depending on the preference of management. However, the well-known types of glasses are important for the server to recognize:

- **Whiskey jigger** or **shot glass**. A very small glass with a capacity of $1\frac{1}{2}$ ounces
- **Highball glass**. A medium-tall, straight-sided glass holding between five and eight ounces
- **Collins glass**. A tall, straight-sided, frosted glass holding about 10 to 12 ounces
- **Old-fashioned glass**. A low, squat glass holding about 5 to 10 ounces
- **Cocktail glass**. A four-ounce funnel-shaped stemmed glass
- **Sour glass**. A four-ounce slender, tulip-shaped glass with a short stem
- **Brandy snifter**. An eight-ounce balloon-shaped glass with a short stem
- **Cordial glass**. A slender-stemmed, tulip-shaped glass holding about one ounce
- **Margarita glass**. A 12- to 20-ounce stemmed glass with a wide bowl

Many cocktails are served "up" or **straight up**, meaning without ice. They are served in a stemmed glass to prevent the guest's hands from warming the beverage. Another term used is **neat**, a drink served without ice or mixer. However, if the guest orders a cocktail **on-the-rocks**, serve the cocktail over cubes of ice in an old-fashioned glass. Some of the glasses used in restaurants are shown in Figure 8-13.

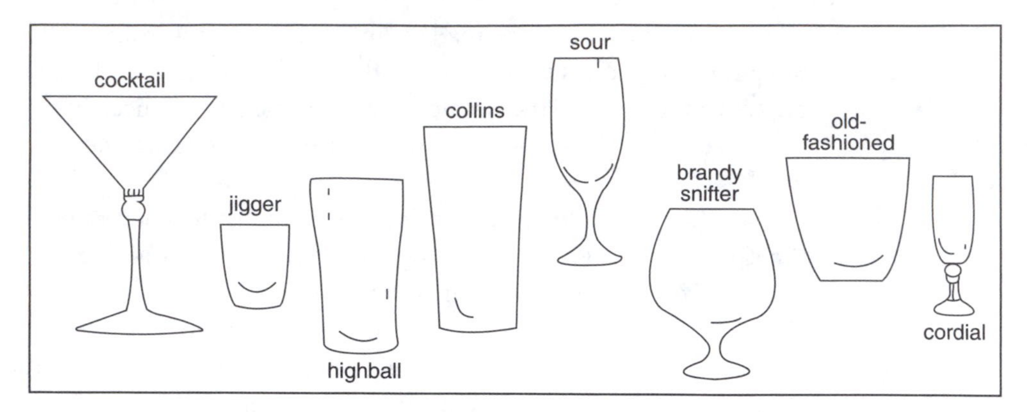

FIGURE 8-13 Examples of Frequently Used Liquor Glasses: An alcoholic beverage or cocktail should always be served in the glass appropriate for the drink. Here are some examples of frequently used liquor glasses.

POPULAR COCKTAILS AND MIXED DRINKS

To serve cocktails or mixed drinks properly, the server must have knowledge of popular drinks (Figure 8-14). Table 8-4 shows the type of glass, ice, ingredients, and garnish for the drinks frequently ordered in restaurants.

PROCEDURE FOR SERVING DRINKS

After the guests are seated and receive their menus, take the cocktail order. Because of the many drink variations, the server must take the order correctly. Many guests request a particular brand of liquor in their drinks, a variation of mixers (e.g., club soda, water, ginger ale), different liquor than traditionally used, or a special garnish.

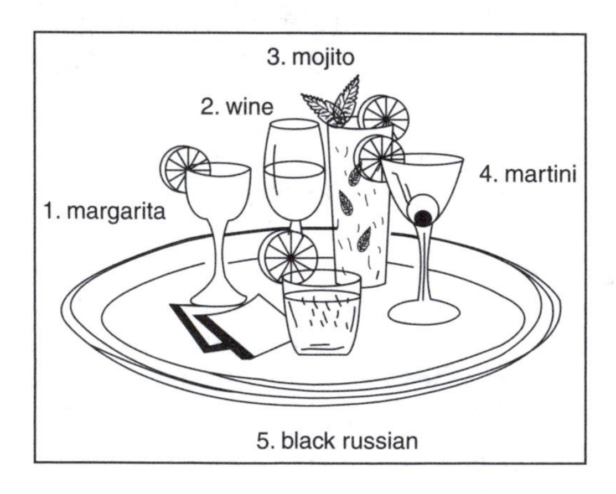

FIGURE 8-14 Popular Drinks on a Tray: A good server should have knowledge of popular drinks such as the ones shown here on a tray.

TABLE 8-4 Popular Alcoholic Drinks

Black Russian
Glass: Old-fashioned
Ice: Mix with ice in a shaker or blender
Ingredients: Coffee liqueur, vodka
Garnish:

Bloody Mary
Glass: Highball
Ice: Pour over cubes in glass
Ingredients: Vodka, tomato juice, hot pepper sauce, Worcestershire sauce
Garnish: Celery stick and lime slice

Cosmopolitan
Glass: Cocktail
Ice: Stir with ice, then strain into glass
Ingredients: Vodka, cranberry juice, dry vermouth
Garnish: Twist of lemon

Long Island Ice Tea
Glass: Collins
Ice: Pour over cubes in glass
Ingredients: Vodka, light rum, gin, tequila, sour mix, splash of Coke
Garnish: Lemon slice

Manhattan (also many variations)
Glass: Cocktail
Ice: Stir with ice, then strain into glass
Ingredients: Whiskey, sweet vermouth, Angostura Bitters
Garnish: Maraschino cherry

Margarita (also many variations)
Glass: Margarita stemmed glass; rim rubbed with lime then dipped in coarse salt
Ice: Shake with ice, then strain into glass
Ingredients: Tequila, triple sec, lime juice
Garnish: Lime slice

Martini (also many variations)
Glass: Cocktail
Ice: Stir with ice, then strain into glass
Ingredients: Gin, dry vermouth
Garnish: Olive or lemon twist

Mojito
Glass: Old-fashioned
Ice: Pour muddled mint, lime, rum, and sugar over crushed ice in glass; add soda
Ingredients: Mint sprigs, lime, light rum, sugar, club soda
Garnish: Mint sprigs

(Continued)

TABLE 8-4 Popular Alcoholic Drinks (Continued)

Pina Colada
Glass: Collins
Ice: Mix with ice in a blender or shaker
Ingredients: Pineapple juice, coconut cream, gold rum
Garnish: Maraschino cherry and pineapple stick

Rob Roy (Scotch Manhattan)
Glass: Cocktail or old-fashioned
Ice: Stir with ice, then strain into cocktail glass, or serve on-the-rocks
Ingredients: Scotch, sweet vermouth, Angostura Bitters to taste
Garnish: Maraschino cherry

Tequila Sunrise
Glass: Collins
Ice: Mix in a blender with ice, then pour into glass. Add grenadine last.
Ingredients: Orange juice, tequila, juice of $1/2$ a lime, grenadine
Garnish: Lime slice

For example, a guest may order a vodka gimlet instead of a gin gimlet or a twist instead of an olive in a dry martini.

After you have properly taken the bar order, submit it to the bartender at the **service bar**, which is an area of the bar for server and bartender use only. You may be required to assist the bartender in preparing the drinks by setting the appropriate glasses on the cocktail tray, icing the glasses when required, adding the mixers, or garnishing the cocktails. Place the drinks on the tray in the order in which you took the orders to help you serve the correct drink to each guest. In some restaurants, the policy is to set up ingredients for highballs, on-the-rocks drinks, and straight drinks at the service bar and to combine the ingredients later at the table in front of the guests.

When drink ingredients are combined in front of the guests at the table, pour the measured liquor into the proper glass, then add the proper amount of mixer designated by the guest. When on-the-rocks drinks are prepared in front of the guests, pour the liquor over the ice in the glass.

When drinks are ready to serve, apply the ordinary rules of serving beverages. Serve from the right with the right hand, and serve women first. Place the drink to the right and slightly below the water glass. Remove drink glasses when empty. Ask the guests whether they would like another cocktail before ordering the meal.

After the guests complete the entrée or dessert and the soiled dishes are cleared, suggest an after-dinner drink such as brandy or a cordial. Serve these drinks in the same manner as you serve other cocktails and mixed drinks.

Nonalcoholic Drinks

As guests become increasingly concerned about nutrition, weight control, and driving while intoxicated, they are shifting their tastes to lighter drinks, such as low- or no-alcohol beers and wines and other beverages without alcohol. One of the biggest innovations in bar drinks has been **nonalcoholic drinks** (see Table 8-5). These drinks are alcohol-free but contain other bar ingredients such as fruit juice

TABLE 8-5 Examples of Nonalcoholic Drinks

Cinderella
Glass: Highball
Ice: Pour over cubes
Ingredients: Club soda, grenadine, lemon juice, orange juice, pineapple juice
Garnish: Pineapple slice and/or orange slice

Iced Coffee or Iced Frappuccino
Glass: Highball
Ice: Blend ingredients with ice
Ingredients: Strong cold coffee, sugar, and milk
Garnish: None

Mango Smoothie
Glass: Specialty glass
Ice: Blend ingredients with ice
Ingredients: Pineapple juice, mango slices, vanilla yogurt
Garnish: None

Shirley Temple
Glass: Highball
Ice: Pour over cubes
Ingredients: Ginger ale, grenadine
Garnish: Maraschino cherry and orange slice

Sparkling Mineral Water
Glass: Highball
Ice: Pour over cubes
Ingredients: Bottled sparkling water
Garnish: Lime or lemon slice

Virgin Mary
Glass: Highball
Ice: Pour over cubes
Ingredients: Tomato or V-8 juice, lemon juice, Worcestershire sauce, Tabasco sauce, celery salt or celery seed, dill, pepper
Garnish: Celery stick and lime slice

and mixers. They may be concocted and garnished to resemble alcoholic drinks, such as coolers, fruit fizzes, frappés, and ice cream–based smoothies. A popular expression for a nonalcoholic drink is a virgin drink, such as a Virgin Mary—a Bloody Mary without vodka. Children's cocktails, such as a Shirley Temple, are always nonalcoholic.

KEY TERMS

Wine
White wines
Pink wines
Red wines
Rosé wines
Blush wines
Body
Bouquet
Dry wine
Varietal name
Generic names
Proprietary names
Vintage
Table wines
Sparkling wines
Fortified wines
Aromatized wines
Wine cooler
Fifths
Wine glass
Wine list
Wine chart
Waiter's corkscrew
Beer
Lager beer
Ale
Pilsner
Munchner
Weisse
Malt liquors
Light beer
Bock beer
Footed pilsner
Stein
Brandy
Whiskey
Vodka
Gin
Rum
Aquavit
Tequila
Cordials
Jigger
Highball glass
Collins glass
Old-fashioned glass
Cocktail glass
Sour glass
Brandy snifter
Cordial glass
Margarita glass
Straight up
Neat
On-the-rocks
Service bar
Nonalcoholic drinks

Review

1. What are the advantages of serving alcoholic beverages in a restaurant?
2. What are the restaurant personnel's responsibilities in regard to serving minors or intoxicated guests?
3. How does blush wine differ from rosé?
4. What is the bouquet of a wine? How does the server get the red wine bouquet to develop?
5. When are dry wines and sweet wines usually served during the meal?
6. What does the vintage tell you about a wine?
7. Why should corked wines be stored horizontally?
8. What is the maximum time that wine should be refrigerated?
9. Why is an ice bucket or wine cooler used in the service of wine?
10. What wines would you suggest with the following foods: lasagna, prime rib, shrimp appetizer, chicken, cheesecake, lobster, cheese soufflé, sirloin steak, Cobb salad?

11. What amount of wine (fifth, split, glass) would you suggest for two people? Four people? Twelve people?
12. Why are wine glasses stemmed?
13. Why should a bottle of wine be handled with care before opening it?
14. What purpose does a wine chart serve?
15. What is the procedure for pouring beer for a guest?
16. What are the distinguishing features of brandy, whiskey, vodka, gin, rum, and cordials?
17. Why is it important to know the glassware used for alcoholic beverages?
18. What are some of the ingredients found in a nonalcoholic cocktail?

Projects

1. Have a group discussion about the effect of alcohol on the body and the server's responsibility toward serving alcohol responsibly. Discuss some difficult situations in terms of serving alcohol in a restaurant.
2. From your restaurant's wine list or any wine list, identify the types of wine listed, such as table, sparkling, fortified, and aromatized. Determine the temperature at which you would serve each wine. Ask a bartender or wine sommelier to answer any questions you might have about the wine list.
3. Attend a short, commentated tasting of wines on a good wine list. Note the qualities of each wine. You can read many books on wine, but nothing replaces firsthand tasting of the wine in question. You as the server are most likely to encourage guests to try the wines you have tasted and liked. At another time, you might also have a commentated tasting of beer.
4. Obtain examples of glassware used for drinks, and list the drinks that are served in each glass.
5. Memorize the table of popular drinks in this book, and learn those drinks popular in your area. Identify the setup, including the glassware, ice, and garnish, as a fellow trainee names each drink.
6. Role-play serving alcoholic beverages. Ask some of the trainees to be guests. Practice suggesting drinks, taking the order, opening and serving table wines and sparkling wines, and serving beer, cocktails, and mixed drinks (including combining drinks at the table).

Case Problem

A Bad Suggestion

You, as a server, make a recommendation to a guest about a wine, and the guest decides to try it. The wine is brought to the table, opened, and, according to custom, a sample is poured for the host. He tastes it but does not like it, saying it is "acidic" instead of "soft" as you described. Answer the following questions:

- Who should take responsibility for the rejected bottle of wine?
- Why did this happen?
- What are the exact words you would use to convey to the guest that you understand his objection to the wine?
- Would you take back the wine?

Definitions of Key Terms from the Text

A

À la carte
Single item priced separately from other foods

Ale
Made with different proportions of beer ingredients and fermented at a higher temperature than lager beer

American service
Food is dished up on the dinner plates in the kitchen

Appetizer
Food that is served first and is offered to arouse the appetite

Aquavit
Scandinavian distilled beverage made the same way as gin but with a caraway flavor

Aromatized wines
Wines lightly fortified and flavored with herbs, spices, and peels

B

Baked
Cooked by dry, continuous heat in an oven

Banquet menus
Preplanned menus for large groups of guests

Banquet service
The menu, number of guests, and time of service are predetermined in advance

Beer
Brewed alcoholic beverage made from fermented barley, malt, hops, yeast, and water

Blackboard menus
Menus written on blackboards (so they can be easily changed)

Blush wines
Pink wine made from one grape variety

Bock beer
Dark beer made in the spring with more body than usual and pleasant bitterness from the hops

Body
Thickness or thinness of wine

Boiled
Cooked in liquid at the boiling temperature of 212°F at sea level

Bouquet
Aroma or fragrance wine emits as it is swirled

Braised
Browned in a small amount of fat and then cooked in a small amount of liquid in a covered pan

Brandy
Liquor made by the distillation of wine or fermented fruit mash

Brandy snifter
An eight-ounce balloon-shaped glass with a short stem

Broiled
Cooked either in a broiler or on a grill

Buffet service
Guests select their meals from an attractive arrangement of food on long serving tables

Busser
An employee of the restaurant who assists the server in serving the menu items and clearing soiled dishware, as well as cleaning and resetting tables

C

Cash drawer
Drawer for making change for guest checks paid in cash

Centers for Disease Control and Prevention (CDC)
Government agency in charge of protecting the health and safety of citizens

Checklist order system
The server selects the food choice from a preprinted list of menu items

Chef de rang
In French service, the experienced server who takes the order, serves the drinks, and prepares some of the food with flourish at the guest's table

Children's menus
Menus with children's favorite foods, child-size portions, and low prices

Cocktail glass
A four-ounce funnel-shaped stemmed glass

Collins glass
Tall, straight-sided glass holding 10 to 12 ounces

Commis de rang
In French service, serves the food as dished up by the Chef de rang

Condiments
Additives used to give flavor and relish to food, such as salt, pepper, ketchup, steak sauce, etc.

Cordial glass
Slender-stemmed, tulip-shaped glass holding about one ounce

Cordials
Sweet, colorful drinks containing at least 2 percent sugar

Cover
Arrangement of china, silverware, napkin, and glassware at each place setting

Crumber
Pencil-length metal tool with a groove in it for sweeping the table clean

Crumbing the table
Sweeping loose food particles into a clean plate with a folded napkin or a crumber

D

Decaffeinated coffee
Coffee with only a trace of caffeine

Dessert table
A self-service attractive display of preportioned desserts

Deuce
Table for two

Dining room manager
A restaurant management employee who directs and coordinates foodservice in the dining room; hires, trains, and supervises employees; and handles budgets, payroll, and purchasing; reports to general manager

Disabled person
One who has a physical or mental impairment

Diversity
Difference or unlikeness, and refers to the fact that each person is unique with regard to race, color, creed, ethnicity, religion, national origin, gender, sexual orientation, disability, age, marital status, socioeconomic status, veteran status, belief, or ideology—to name just a few dimensions

Drink runner
A restaurant employee who brings drinks from the bar

E

Early bird menus
Menus for guests who are willing to dine before a certain time

English service
The food is brought from the kitchen on platters and placed before the host to serve

Entrées
In American service, the main part of a meal, including such items as steaks, seafood, poultry, etc.

Expeditor
A restaurant employee who assembles and brings orders to the dining tables for servers

F

Family-style service
Food is cooked in the kitchen and then dished into large bowls and platters and placed in the center of the tables

Fifth
Bottle of wine containing 25.4 ounces or 750 ml

Finger bowls
Bowls of warm water with rose petals or lemon slices in them to rinse the fingers at the table

First aid
Treatment for minor injuries

First Aid for Conscious Choking Emergencies
Recommended procedure to use when responding to a choking incident

Food allergy
Sensitivity to food ingredients that may cause an adverse physical reaction

Foodborne disease outbreak
An incident in which two or more persons experience a similar illness after ingestion of a common food

Foodborne illnesses
Infection caused by foodborne pathogens when eating unsafe food

Footed pilsner
Tall, narrow, seven-ounce beer glass with a short stem

Fortified wines
Wines combined with brandy to increase alcohol content

French service
Formal type of service employing two servers working together to serve the meal and cooking or completing food preparation at a side table in front of guests

Fried
Cooked in hot fat

G

Generic names
Wines that are named after the geographic region where the grapes are grown

Gin
Redistilled liquor employing the juniper berry as the principal flavoring agent

Gratuity
See Tip

Grilled
Cooked over direct heat

Grooming
The process of making your appearance neat and attractive

Guéridon
Cart used for tableside cooking

Guest-check order system
Server handwrites the order directly on the guest-check form

Guest paging device
Wireless, remote, guest call system that allows the host to signal the party of guests when their table is ready

H

Handheld computer order system
Order is taken directly on a handheld computer

Handheld order terminal
Small, portable, wireless POS system used by servers to take orders from guests at their tables

Harassment
Disturb, worry, unnerve, or torment by continuous small attacks

Hazard Analysis Critical Control Point (HACCP)
Part of the Food Code; a food safety system to prevent outbreaks of foodborne illness

Head Server
Server who supervises and coordinates dining room employees for a section of the dining room, and may greet, seat, and serve guests; reports to host or maître d'hôtel

Highball glass
Medium-tall, straight-sided glass holding five to eight ounces

Hospitality
Creating a pleasant dining experience for your guests with various unexpected small gestures

Host
A restaurant employee who schedules shifts and assigns stations, holds daily meetings with staff, controls the flow of seating in the dining room, and greets, seats, and provides menus to guests; reports to dining room manager

Hygiene
Practices that promote personal cleanliness and good health

J

Jigger
Very small glass with a capacity of $1\frac{1}{2}$ ounces

K

Kosher foods
Foods permitted to be eaten by people of the Jewish faith who observe kosher dietary law

L

Lager beer
Pale, aged, effervescent brew made with the addition of corn and rice to give it its light body

Light beer
Lager beer brewed to reduce the number of carbohydrates and calories

M

Magnetic card reader
Device attached to or part of a computer terminal that reads numbers on a card pulled through a slot

Maître d'hôtel
A restaurant employee who schedules shifts and assigns stations, holds daily meetings with staff, controls the flow of seating in the dining room, and greets, seats, and provides menus to guests; reports to dining room manager (title generally used in fine-dining establishments and often abbreviated simply as maître d)

Malt liquors
Lager beers with an alcoholic content of 4 to 8 percent

Margarita glass
A 12- to 20-ounce stemmed glass with a wide bowl

Material Safety Data Sheets
Cleaning product safety information in case a chemical is inhaled or ingested

Meal accompaniments
Condiments, decorative garnishes, and foods that complement the entrée

Menu
Food offerings of a restaurant and the actual printed or readable list of those foods

Mise en place
Sidework; all duties the server performs that are not directly related to serving the guests

Munchner
Munich beer with a deep brown color that is slightly sweet and has a strong malt flavor

N

Neat
Drink served without ice or mixer

Nonalcoholic drinks
Alcohol-free drinks that contain other bar ingredients, such as fruit juice and mixers

Notepad order system
The server writes the order on a blank pad of paper

O

Old-fashioned glass
Low, squat glass holding five to ten ounces

On-the-rocks
Served over cubes of ice in an old-fashioned glass

Outbreak
Incidence of foodborne illness when two or more people become ill after eating the same food

Organization chart
Arrangement of jobs in a restaurant diagrammed on paper

Oyster bar
Buffet featuring oysters on the half shell and various seafood and mustard sauces that complement the oysters

P

Pager
Electronic device that signals the server by beeping or vibrating

Pan-broiled
Cooked in a heavy frying pan over direct heat, using little or no fat

Pareve
Foods containing neither meat nor dairy products

Pilsner
Lager beer with a pale, golden color and a strong hops flavor

Pink wines (rosé and blush)
Made by separating the skins of the grapes soon after fermentation has begun

Poached
Simmered in enough liquid to cover the food

Preparation time
The time required to cook and dish up a food item on the menu

Printer
Imprints information on food and beverage orders, guest checks, receipts, and management reports

Proprietary names
Brand names adopted by the bottler for sales purposes

R

Receptionist/reservationist
One who handles reservations

Rechaud
Small spirit stove used in the dining room to keep the food warm

Red wines
Made by allowing the skins of the grapes to remain in the tank

Restaurant manager
Responsible for the efficiency and profitability of the restaurant operation; reports to restaurant owner

Restaurant point-of-service (POS) systems
Computer systems that combine hardware and software to automate restaurant transactions and functions

Roasted
Cooked uncovered without water added, usually in an oven

Rosé wines
Pink table wines that share the characteristics of both red and white table wines

Rum
Distilled beverage made from sugar cane by-products

Russian service
Elegant service that employs the use of heavy silver serviceware and differs from French service in that it uses only one server and the food is fully prepared in the kitchen

S

Safety
Freedom from harm or danger

Sanitation
Process of working out ways to improve health conditions

Salad bar
A self-service concept allowing the guests to prepare their own salads

Sautéed
Browned or cooked in a small amount of hot fat

Servers
Also referred to as waitstaff; employees who create a pleasant experience in the dining room for guests by taking care of guests' needs

Server terminal
Touch screen monitor found near the dining room or bar

Service
Filling the needs, wants, and desires of guests

Service bar
Area of the bar for server and bartender use only

Serving
Bringing the food and beverage order to the table and attending to the guests' needs

Sidestand
Storage and service unit, sometimes with a computer terminal, located close to serving areas

Sidework
Also called mise en place, and includes all of the duties the server performs other than those related to serving the guests

Silencer
A pad or second tablecloth placed beneath the top cloth to give the table a better appearance and to soften the sound of tableware

Simmered
Cooked in liquid over low heat just below the boiling point

Smorgasbord
A buffet featuring a large selection of Scandinavian food, such as cheese and herring

Sneeze guard
A clear, protective panel mounted above a self-service food station

Software
Application or instructions to run the computers

Sour glass
A four-ounce slender, tulip-shaped glass with a short stem

Sparkling wines
Wines containing carbon dioxide often used for special occasions

Special
Chef's specialty, a regional dish, or a seasonal food in ample supply

Station
Section of the dining room assigned as a work area to a server

Steamed
Cooked in steam with or without pressure

Stein
Heavy beer glass with a handle

Stewed
Simmered slowly in enough liquid to cover the food

Straight up
Without ice

Suggestive selling
Suggesting additional food items to add to the pleasure of the meal and increase the size of the check, restaurant profits, and tip

T

Table d'hôte
Full-course meal priced as a unit; sometimes called prix fixe (fixed price in French)

Table management software
Tracks reservations and seating, as well as which server is responsible for which tables

Table tent menus
Small menus designed to stand vertically

Table wines
Unfortified white and red wines that accompany a meal

Taking reservations
Promising a table to guests who call or e-mail in advance

Tea
Aromatic beverage made by infusing tea leaves in boiling water

Tea sachet
Individually bagged portion of loose tea

Teamwork
Cooperating and working together with coworkers and the supervisor of the dining room to serve the public

Tequila
Mexican spirit distilled from the juice of the agave

Tip (Gratuity)
Monetary reward for courteous and efficient service

Tornado warning
A funnel has actually been sighted in your area

Tornado watch
Conditions are favorable for tornadoes to develop

U

Underliner
A small plate that underlines another container of food

Uniform
Garment that identifies the occupation of the wearer

Universal Precautions
Procedure to protect people who may be exposed to blood or other body fluids due to contact with an ill or injured person

U.S. Food and Drug Administration (FDA) Food Code
Model of sanitation rules and regulations that every state adopts and implements as its legislature requires

V

Varietal name
Name of the primary grape used in making the wine

Vintage
Wine from a grape crop or harvest of a good year

Violence
Rough or harmful action or treatment

Vodka
Distilled from fermented grain mash but not aged and is colorless, without aroma or flavor

W

Waiter's corkscrew
Wine opener for wine bottle sealed with a cork

Waitstaff
Also referred to as servers; restaurant employees who create a pleasant dining experience for guests in a restaurant by taking care of guests' needs

Weisse
White German beer made from wheat

Whiskey
Liquor distilled from fermented grain and aged in wooden barrels

White wines
Made by fermenting the grape juice without skins

Wine
Beverage made from fermented grape juice and containing 10 to 15 percent alcohol

Wine chart
Describes available wines and pairs them with menu items

Wine cooler
Ice bucket used to maintain the proper temperature of wine at the table

Wine glass
Thin, steamed, tulip-shaped glass, free of decoration and color

Wine list
Wine menu

Write-on handheld order terminal
Works like a pencil and pad; servers use a stylus to bring up all possible menu items that can be selected

Resource B

Definitions of Menu and Service Terms

A

À la (ah-lah)
Prepared in a particular manner

À la king (ah-lah-KING)
Food served in white sauce with mushrooms, green peppers, and pimientos

À la mode (ah-lah-MODE)
Usually dessert with ice cream; sometimes style of the day

Alfresco dining (al-FRES-ko)
Serving food in the fresh air; outdoors

Allemande (al-leh-mahnd)
Wine sauce with butter, egg yolk, and ketchup, which give it a yellow color when combined

Amandine (ah-mahn-DEEN)
With almonds added

Ambrosia (am-BRO-zha)
Fruit dessert consisting of oranges, bananas, and shredded coconut

Americano
Espresso diluted with steaming water; a weaker espresso

Anglaise (ahng-glayz)
Cooked in water or stock

Antipasto (ahn-tee-PAHS-toe)
Italian appetizer that includes raw vegetables, fish, and meat

Aspic (as-pick)
Clear vegetable, meat, fish, or poultry jelly

Au gratin (oh-GRAH-tin)
Prepared with a sauce and baked with a topping of bread crumbs and cheese

Au jus (oh-ZHUE)
With unthickened natural juices

B

Bagel (BAY-gul)
Ring-shaped roll with a tough, chewy texture, made from plain yeast dough that is dropped briefly into nearly boiling water and then baked

Baked Alaska
Ice cream on cake, covered with meringue and baked in an oven until the meringue browns

Bardé (bar-DAY)
Covered with pork or bacon slices

Basted
Stock drippings, or fat moistening or spooned over food while cooking

Battered
Covered with a mixture of flour and liquid of a consistency that can be stirred

Bavarian or Bavarois (bav-ar-waz)
Cream gelatin with whipped cream folded into it as it begins to stiffen

Béarnaise sauce (bair-NAZE)
Sauce similar to hollandaise and containing tarragon

Béchamel (bay-shah-MEL)
Rich cream or white sauce

Beef à la mode (Beef ah-lah-MODE)
Larded piece of beef cooked slowly in water with vegetables; similar to braised beef

Beef Stroganoff
Sautéed tenderloin of beef with a sour cream sauce

Beignet (ben-YAY)
Holeless donut usually served with café au lait

Belevue (bel-vue)
Food enclosed in aspic that can be seen

Benedictine (ben-eh-DIK-teen)
Liqueur made in Europe

Beurre noir (burr NWAHR)
Butter cooked to a dark brown, to which capers and a dash of vinegar are added

Biscotte (bis-kot)
Biscuit or rusk

Bisque (bisk)
Rich cream soup, often made with seafood

Blanquette (blang-KET)
White stew, often made with veal

Blinis (BLEE-neez)
Russian buckwheat pancakes, often served with caviar

Blintzes (BLIHNTS)
Ricotta cheese rolled in crêpes and fried in pan with butter

Blue cheese
Similar to Roquefort in appearance but made of cow's milk instead of sheep's milk

Boeuf (buhf)
Beef

Bombé (bom-BAY) or Bombé glacé (gloss-AY)
Frozen dessert that is a combination of two or more frozen mixtures packed in a round mold

Bonne femme (bun-FAM)
Simple homestyle soups, stews, etc.

Bordelaise (bohr-dih-LAYZ)
Brown sauce made with butter or marrow fat, meat stock, bay leaf, onions, carrots, red or white wine, and seasonings

Borsch or Borscht (borsh or borsht)
Russian or Polish soup made with beets

Bouillabaisse (boo-yah-BAYS)
Several varieties of fish fixed as a stew

Bouillon (BULL-yon)
Meat broth

Bourgeois (boor-ZHWAZ)
Often means served with vegetables

Bourguignonne (boor-gee-NYON)
Usually beef braised in red wine and garnished with mushrooms and onions

Breaded
Rolled in bread crumbs or other breading agent before cooking

Brioche (bre-OSH)
Lightly sweetened rich bread

Broche (broash)
Skewer or spit for roasting

Brochette (broe-SHET)
Meat broiled on a skewer

Brunoise (brun-WAHZ)
Finely diced

C

Cacciatore (caht-chah-TORE-ee)
Sauce containing tomatoes, onions, garlic, and herb spices; wine may be added

Caesar salad (SEE-zer)
Green salad with anchovies, croutons, Parmesan cheese, coddled eggs, and garlic

Café (kaf-AY)
Coffee, coffee house, or restaurant

Cajun (KAY-juhn)
Made with a dark roux with fat and spices, celery, green peppers, and onions

Canadian bacon
Smoked loin of trimmed, lean pork

Canapé (kan-a-PAY)
Spicy food mixture spread on crackers, toast, or bread

Cannelloni (kan-a-LO-nee)
Pasta or crêpe stuffed with cheese or meat and served with tomato or meat sauce

Capon (kay-POHN)
Castrated poultry noted for tenderness and flavor

Cappuccino (CA-pa-CHEE-no)
Equal parts of brewed espresso, steamed milk, and froth from the steamed milk, often served with sugar

Carte du jour (kart du ZHOOR)
Menu of the day

Casaba (kah-SAH-bah)
Large, oval melon with yellow skin and white meat

Caviar (ka-vee-AR)
Gray and black eggs or roe of fish; red eggs if from salmon

Cèpes (sep)
Particular species of mushrooms

Champignons (shahm-peen-YONE)
Mushrooms

Chanterelles (shahn-teh-REL)
Species of mushrooms

Chantilly cream (shahn-tilly)
Vanilla whipped cream

Chantilly sauce (shahn-tilly)
Hollandaise sauce with whipped cream

Chartreuse (shar-truhz)
Food with a hidden filling; also a certain liqueur

Chateaubriand (shah-TOE-bree-ahn)
Thick tenderloin steak

Chaud (sho)
Hot

Chef
Person in charge of food preparation

Chef's salad
Green salad topped with strips of ham, cheese, and chicken

Chemise (sheh-MEEZ) or En chemise
Skins on; often refers to potatoes

Chiffonade (shif-eh-NADE)
Shredded vegetables sprinkled on salads or soups

Chimichanga (chee-mee-CHAN-gah)
Deep-fried burrito

Cloche (kloash)
Dish cover

Club sandwich
Sandwich made with three layers of bread or toast and filled with chicken, bacon, and tomato

Cobbler
Deep dish fruit pie, or drink of wine or liquor with sugar, sliced fruit and mint

Cocktail
Appetizer or alcoholic drink

Compote (KOM-pote)
Stewed fruit combination

Confiture (KON-fee-chure)
Fruit jam or preserves

Consommé (kon-so-MAY)
Clear broth

Continental breakfast
Assorted juices, coffee, sweet rolls, and other food items already prepared

Course
Part of a meal served at one time

Crème de menthe (krem-deh-MENTH)
Peppermint cordial

Creole (KRE-ole)
Made with tomatoes, onions, peppers, and seasoning

Crêpes suzette (krape su-ZET)
Thin, fried pancakes covered with a sauce of liqueur and served flaming

Croissant (krwa-SAN)
Crescent-shaped roll; sometimes a confectionery

Croutons (KROO-tahns)
Small cubes of bread fried or baked until brown

Cuisine (kwee-ZEEN)
Cookery; also kitchen

Curry
East Indian type of stew made with curry powder

D

Dejeuner (DAY-zhuh-nay)
Breakfast or lunch

De la maison (de lah ma-ZON)
House specialty

Demi (de-mee)
Half

Demitasse (DEM-ee-tass)
Small cup of strong coffee

Diable (de-OBBEL)
Deviled

Diner (de-nay) (French)
Dinner or to dine

Drawn butter
Melted butter

Duchesses potatoes (DUCH-ihs)
Potatoes mashed with eggs and forced through a pastry tube

Dugiére (doog-LAIR)
Tomatoes are used

Du jour (doo ZHURE)
Of the day

Dusted
Sprinkled with sugar or flour

E

Éclair (ay-KLAIR)
Oblong cream puff filled with custard and iced

Eggplant
Large, purple-skinned, pear-shaped vegetable

Eggs Benedict
Poached egg served with ham or tongue with hollandaise sauce on toasted muffin

Emincé (ay-man-SAY)
Cut finely

En casserole (ahn-KAHS-eh-ROLE)
Food served in the dish it was baked in; also (casserole) oven-safe dish with cover

Enchilada (en-chuh-LAH-dah)
Mexican dish consisting of meat or cheese rolled in a tortilla covered with a peppery tomato sauce

En coquille (ahn-koh-KEE)
In the shell, such as oysters on the half shell

Entrée (ON-tray)
Main course of a meal in American service

Escargot (es-kar-GO)
Snail

Espagnole (ays-pah-NYOLE)
Brown sauce

Espresso (es-PRES-o)
Very strong black coffee made of coffee beans roasted black and brewed under steam pressure in an espresso machine

Ethnic fusion dish
Menu item in which ingredients from two or more ethnic cuisines are combined to produce an individual item

F

Fajitas (fah-HEE-tuhs)
Mixture of beef or chicken, onions, and green peppers served sizzling hot on an iron skillet with tortillas

Fanchonette (fan-sho-NET)
Tiny pie or tart covered with meringue

Farce (farse)
Stuffing or forcemeat

Farci (far-SEE)
Stuffed

Farina (fuh-REE-nuh)
Coarsely ground inner portion of hard wheat

Farinaceous
Made with meal of flour

Femiére (fe-mee-AIR)
Made with diced potatoes, carrots, onions, turnips, celery, and cabbage; also farmer-style

Fettuccine (feht-tuh-CHEE-nee)
Square or convex long pasta

Filet mignon (fih-LAY meen-YONE)
Beef tenderloin

Fillet (fih-LAY)
Boneless cut of meat or fish

Finnan haddie
Smoked haddock

Flambé (flahm-BAY)
Served with flaming liqueur

Flenron (flen-rohn)
Baked, crescent-shaped puff pastry used as a garnish, often for fish or white sauce

Florentine (FLOR-ahn-teen)
With spinach

Foie gras (fwa gra)
Fat liver; most often liver of fat geese

Fondue (fon-DUE)
Melted or blended

Forcemeat
Chopped meat with seasoning used for stuffing

Franconia
Browned; usually potatoes browned with a roast

Frappé (frap-PAY)
Beaten and iced drink

Fromage (froe-MAHZH)
Cheese

G

Garbanzo (gar-BON-zo)
Chickpea

Garnish or Garniture (GAR-nee-ture)
Decorate; food item used to decorate

Gefilte fish (ge-FIL-teh)
Fish dumpling

Gherkins
Pickled, small, young cucumbers

Giblets (JIHB-lihts)
Poultry, liver, heart, and trimmings

Glacé (glah-SAY)
Glossy or semitransparent coating

Gnocchi (NAH-kee)
Italian dumpling

Gourmet (goor-MAY)
Expert connoisseur of food and drinks

Gratin (grah-tan) or Gratinée (GRA-tin-ay)
Dusted or sprinkled with cheese or buttered crumbs and baked brown

Gruyère (grae-YER)
Swiss cheese that tastes tarter and has smaller holes than regular Swiss cheese

Guava (gwa-va)
Apple- or pear-shaped tropical fruit with an acidic, sweet flavor, made into jams and jellies

Gumbo (GUHM-boh)
Soup or stew, often made of seafood or chicken, okra, green peppers, and tomatoes

H

Haché (hah-shay)
Chopped or minced

Hasenpfeffer (HOSS-en-feffer)
Rabbit stew

Hollandaise (hah-len-DAZE)
Sauce made with egg yolk, butter, and lemon juice

Hors d'oeuvres (or-DURV)
Small appetizers

Huevos rancheros (WAY-vohs rahn-CHE-rohs)
Eggs with salsa served with tortilla and frijoles refritos

Hush puppies
Southern deep-fried cornmeal cakes

I

Indian pudding
Slowly baked dessert made of cornmeal, milk, brown sugar, eggs, and raisins

Italienne (e-tal-ee-EN)
Italian style

J

Jambalaya (juhm-buh-LI-yah)
Ragout or hash usually with ham and rice

Jardiniere (zhar-dee-NYAIR)
With vegetables

Johnnycake
Cornbread made from yellow cornmeal, eggs, and milk

Julienne (joo-lee-EN)
Thin strips of food

Jus (zhuse)
Juices from meat

K

Kabob (ka-BOB)
Cubes of meat and other foods cooked on a skewer

Kipper
Method of preserving herring, salmon, and other fish

Kosher (KOH-sher)
Jewish biblical term used to describe foods that are permitted to be eaten by people of the Jewish faith. Some feel these foods represent quality, cleanliness, and purity

Kuchen (ku-khen)
Cake

L

Lait (lay)
Milk

Laitue (lay-tu)
Lettuce

Langouste (lahn-goost)
Crawfish

Lasagne (luh-ZAHN-yah)
Frilly or curly-edged pasta

Latke (LAHT-kuh)
Potato pancakes served as a side or as a meal

Latte (LAH-tay)
Coffee beverage that is predominantly cream

Lebkuchen (leb-ku-khen)
German sweet cakes or honey cakes

Leek
Small, onionlike vegetable

Legume (lay-GEWM)
Vegetable; also such foods as peas, beans, and lentils

Limpa (LIHM-puh)
Swedish rye bread

Linguine (lin-GWEE-nee)
Square or convex long pasta

Lox
Smoked salmon

Lyonnaise (lye-a-NAYZ)
Sliced or chopped food fried in butter with onions

M

Madrilene (mah-dreh-LAIN)
Clear consommé with tomato seasoning, served hot or jellied

Maître d'hôtel (MAY-treh doe-TEL)
Head of catering department; head of foodservice

Maître d'hôtel, à la (MAY-treh doe-TEL, ah-LAH)
Yellow sauce; butter sauce with lemon juice and parsley

Manhattan clam chowder
Made with tomatoes, vegetables, and quahog clams

Manicotti
Pasta tubes, usually ridged

Maraschino (mah-rahs-KEE-no)
Italian cherry cordial; also cherries

Marengo (muh-RENG-goh)
Sautéed veal or chicken with tomatoes, mushrooms, olives, and olive oil

Marinade (mar-eh-NADE)
French sauce used to tenderize meats and vegetables

Marsala (mahr-SAH-lah)
Pale golden, semidry wine from Sicily

Matelote (MAT-eh-lo)
Fish stewed with onions and wine; also fish stews

Medallion (meh-DAL-yuhn)
Small round or oval serving of food; often meat fillets

Melba toast
Thin slices of even dried toast

Menthe, crème de (krem-deh-MENTH)
Peppermint cordial

Meringue (meh-RANG)
Paste of egg whites and sugar, souffléd

Meunière (men-YARE)
Fish dipped in flour, sautéed in butter, and served with brown butter, lemon, and parsley

Milanese (mee-lan-AYZ)
Garnish consisting of julienne of ham, mushroom, tongue, and truffles

Mince (MIHNS)
Chop finely

Minestrone (min-a-stro-nee)
Macaroni and cheese product in a vegetable soup

Mixed grill
Three kinds of meat or fish broiled together and served on one plate

Mocha (MOE-ka)
Coffee and chocolate mixed together

Mocktails
Alcohol free versions of popular cocktails

Mongol soup
Soup made with tomatoes, split peas, and julienne vegetables

Mornay (mor-NAY)
White sauce with cheese

Mortadella (mohr-tuh-DEHL-uh)
Italian pork and beef sausage

Mostaccioli (MOS-ta-kee-O-lee)
Round, hollow pasta that is smooth or ribbed

Mousse (moose)
Chilled dessert of whipped cream, gelatin, and flavoring

Mozzarella (mot-za-REL-a)
Soft Italian cheese

Muffaletta (moof-fuh-LEHT-tuh)
Circular Italian cold cut sandwich with olives

Mulligatawny (muhl-ih-guh-TAW-nee)
Thick Indian soup seasoned with curry

Mushroom sauce
Sauce made with fat, flour stock, sliced mushrooms, seasoning, and wine

N

Napoleon (nuh-POH-lee-uhn)
Layered oblong pastry with custard, cream, or jam filling

Neapolitan (nee-uh-PAHL-uh-tuhn)
Dessert of two to four kinds of ice cream, ice, or gelatin of different colors

Nesselrode pudding (NEHS-uhl-rohd)
Frozen dessert made with custard, chestnuts, fruit, and cream

Newburg
Creamed dish made with seafood and egg yolk and flavored with sherry

Normande (nor-MAND)
Smooth, delicate mixture containing whipped cream

O

Okra (OH-kruh)
Vegetable pods often used in soups and gumbos

Omelet or Omelette (AHM-leht)
Beaten egg mixture that is cooked and filled with foods such as cheese or meats

Oysters, bluepoints
Oysters from the Atlantic Coast

P

Panache (pah-NASH)
Mixture of several kinds of feathers, fruits, and vegetables fixed decoratively

Pané (Pan-ay)
Breaded

Papaya (puh-PI-yuh)
Tropical fruit

Parboiled
Boiled until partially cooked

Parfait (par-FAY)
Ice cream, fruit, and whipped cream in tall, slender-stemmed glasses

Parisienne (pa-ree-zee-EN) potatoes
Potatoes shaped with a small round scoop; can be browned, steamed, or boiled

Parmesan (PAR-meh-zan)
Grated, hard, sharp cheese used for toppings, soups, and soufflés

Pastrami (pa-STRA-me)
Beef cured with spices

Paysanne (pay-sahn)
Vegetables cut into shapes

Petit (peh-TEE)
Small

Petit dejeuner (peh-TEE day-zhu-NAY)
Breakfast

Petite marmite (peh-TEET mahr-MEET)
Consommé with chicken, beef, and vegetables; also an earthenware pot

Petits fours (peh-tee-FOR)
Small cakes and cookies, usually served with desserts

Pièce de résistance (pee-es de ray-ZEE-stahns)
Main dish

Pilaf or Pilaú (pee-LOH)
Sautéed with onions and stock; also with meats and vegetables

Piquant (pee-KAHN or pee-KAHNT)
Highly seasoned

Poisson (pwah-sone)
Fish

Polonaise (po-lo-NAYZ)
Garnish of bread crumbs browned in butter, chopped hard-cooked egg, beurre noir, and chopped parsley

Pommes de terre (pum de TAIR)
Potatoes (French; fruit of the earth)

Popovers
Quick, individual, puffed up butter rolls made of milk, flour, and eggs

Postum
Coffee substitute made of cereal

Potage (po-TAHZH)
Soup

Pot-au-feu (pot-oh-FUH)
Boiled meats and assorted vegetables with meat broth

Potpourri (po-poo-REE)
Mixture

Poulet (poo-lay)
Chicken

Prosciutto (pro-SHOO-toe)
Dry-cured, spiced ham

Purée (pew-RAY)
Paste or pulp of fruit or vegetables; also thick soup

Q

Quahog (ko-hog)
Round clams from the Atlantic coast

Quenelles (keh-NEL)
Dumplings

Quesadilla (keh-sah-DEE-yah)
Tortillas layered with cheese served hot

Quiche (keesh)
Combination of cream, eggs, Swiss cheese, and other ingredients baked in a prebaked pie shell

R

Ragout (rah-GOO)
Thick stew

Ramekin (RAM-ih-kihn)
Individual portion of some food baked in a halting dish, often topped with cheese and bread crumbs; also small baking dish

Remoulade (ray-muh-LAHD)
Tart flavored mayonnaise used as a dressing for chilled shellfish

Rigatoni
Cylindrical pasta, either smooth or ribbed

Rissole (ris-SOL)
Browned; also a small turnover

Rissole potatoes or Pommes risoleés (ris-o-LAY)
Potatoes cut into egg shapes, browned, and finished in an oven

Riz (ree)
Rice

Romaine (ro-MAIN)
Narrow, long, crisp-leaved lettuce with light-colored inner leaves

Roquefort cheese (ROKE-furt)
Semihard white cheese speckled with mold and made only in Roquefort, France

Roulade (roo-LAHD)
Rolled thin piece of meat, with or without stuffing, that is braised or sautéed

Russian dressing
Salad dressing of mayonnaise, lemon juice, chili sauce, Worcestershire sauce, and pimiento

S

Sashimi (sah-SHEE-mee)
Raw fish that is sliced and served with condiments such as shredded radish, gingerroot, wasabi, and soy sauce

Sauerbraten (SOW-uhr-brah-tihn)
Marinated beef roast served with noodles, boiled potatoes, or dumplings

Schnitzel (SHNIHT-suhl)
Cutlet breaded and fried

Schaum torte (schoum tort)
Foam cake made of meringue and crushed fruit

Serviette
Table napkin

Shad
Type of herring

Shallot (SHAL-uht)
Type of onion

Shirred eggs (SHERD)
Eggs baked in a shallow dish

Shoestring potatoes
Potatoes cut very thin and French fried

Skewer (SKYOO-uhr)
Meat, poultry, or vegetables fastened on a long pin or thin wooden stick during cooking

Sole
Flat whitefish

Sommelier (so-meh-LYAY)
Wine steward

Soufflé (soo-FLAY)
Baked dish made from beaten egg whites combined with egg yolks and various other ingredients, such as cheese, spinach, chicken, or chocolate

Spinach (SPIHN-ihch) lasagne or noodles
Lasagne or noodles that are green because of their spinach content

Sports bar
Bar where alcoholic beverages and food are served and large television sets feature sporting events

Stir fry
To stir very fast while frying in a little oil or fat

Sushi (SOO-shee)
Boiled rice with rice vinegar; **Nigir sushi** is thin slices of raw fish wrapped around a rice filling

T

Tartar sauce (TAHR-tuhr)
Sauce for seafood made of mayonnaise and pickle relish

Timbales (TIM-bels)
Little pastry shells filled with a mixture of chicken, seafood, cheese, fish, or vegetables

Tortiglioni
Cylindrical pasta that can be smooth or ribbed

Tortillas (tor-TEE-yas)
Mexican corn pancakes

Tournedos (toor-nuh-DOZE)
Small tenderloin steaks

Tostada
Fried corn tortilla, with cheese or guacamole

Truffles (TRUHF-uh)
Mushroomlike fungi grown underground

Tutti-frutti (too-tee-FROO-tee)
Fruit mixture, as in ice cream

V

Velouté (vel-oo-TAY)
Cream soup or a thick, creamy sauce

Vermicelli
Thin spaghetti

Vichyssoise (vee-shee-SWAZ)
Cold potato and leek soup

Vinaigrette (vin-eh-GRET)
Dressing made with oil, vinegar, and herbs

W

Waldorf salad (WAWL-dorf)
Salad made with a mixture of apples, celery, nuts, and mayonnaise

Wiener schnitzel (VEE-ner-shnit-sel)
Breaded veal cutlet served with lemon

Wonton (WAHN-than)
Noodle dough stuffed with ground chicken or pork, often added to Chinese soups

Y

Yorkshire pudding (YORK-sheer)
Baked egg and flour mixture served with roast beef

Z

Zucchini (zoo-KEE-nee)
Italian summer squash

Recommended Resources for Further Information

Web Sites

Allergens
www.foodallergy.org
www.cfsan.fda.gov

American Red Cross
www.redcross.org

Centers for Disease Control and Prevention (CDC)
www.cdc.gov

Food Management Magazine
www.food-management.com

Guest Paging Systems
www.pager.net

Leading Wine-Producing Countries
www.pages.drexel.edu
www.cellarnotes.net
www.wine.morefocus.com

Lettuce Entertain You Enterprises, Inc.
www.leye.com (Wildfire Restaurant)

Micros Computer Systems
www.micros.com

National Highway Traffic Safety Administration (NHTSA)
www.nhtsa.dot.gov

National Oceanic and Atmospheric Administration
www.noaa.gov

National Restaurant Association
www.restaurant.org

National Restaurant Association Education Foundation
www.nraef.org
www.servsafe.com

Occupational Safety and Health Administration (OSHA) Part of the U.S. Department of Labor

www.osha.gov
www.osha-safety-training.net

Restaurant Reservations and Table Management
www.opentable.com

U.S. Food and Drug Administration (FDA) Food Code 2005
www.cfsan.fda.gov

Wine Serving Temperatures
www.lovetoknow.com

Books and Pamphlets

Department of Transportation (U.S.), National Highway Traffic Safety Administration (NHTSA). *Traffic Safety Facts 2005: Alcohol.* Washington, DC: NHTSA, 2006 (cited 2006, Oct. 3).

Goldfarb, Sylvia. *Allergy Relief*. New York: Penguin Putnam, 2000.

Herbst, Sharon Tyler. *Food Lover's Companion.* New York: Barron's Educational Series, 2001.

Indiana Diet Committee of the Indiana Dietetic Association. *Indiana Diet Manual*, 7th Ed. Elberfeld, IN: IDA, 2006.

Indiana State Department of Health. *Retail Food Establishment Sanitation Requirements.* Indianapolis, IN: Indiana State Department of Health, 2000.

Kotschevar, Lendal H., and Valentino Luciani. *Presenting Service*, 2nd Ed. Hoboken, NJ: John Wiley & Sons, 2007.

Lipkowitz, Myron A., and Tova Navarra. *Allergies A to Z*, 2nd Ed. New York: Checkmark Books, 2001.

McCarthy, Ed, and Mary Ewing-Mulligan. *Wine for Dummies.* Foster City, CA: IDG Books, 1995.

Muth, Annemarie S. *Allergies Sourcebook*, 2nd Ed. Detroit, MI: Omnigraphics, 2002.

Richardson, Brenda. Indiana Dietetic Association. *Indiana Diet Manual.* Elberfeld, IN: IDA, 2006.

Sullivan, Jim. *Tips for Better Tips.* Appleton, WI: The Coca-Cola Company, 2006.

Walker, Georgianna Walker. *Pocket Resource for Nutrition Assessment*. Chicago: American Dietetic Association, 2005.

Index